Automated Instrumentation for Radioimmunoassay

Editor

Lemuel J. Bowie, Ph.D.
Director
Division of Clinical Biochemistry
Evanston Hospital
Evanston, Illinois
and
Assistant Professor of Clinical Pathology
Northwestern University Medical School
Chicago, Illinois

CRC Press
Taylor & Francis Group
Boca Raton London New York

CRC Press is an imprint of the
Taylor & Francis Group, an informa business

First published 1980 by CRC Press
Taylor & Francis Group
6000 Broken Sound Parkway NW, Suite 300
Boca Raton, FL 33487-2742

Reissued 2018 by CRC Press

Library of Congress Cataloging in Publication Data

Main entry under title:

Automated instrumentation for radioimmunoassay.

Bibliography: p.
Includes index.
1. Radioimmunoassay—Automation. I. Bowie,

Lemuel J.
QP519.9.R3A97 616.07'56 79-17539
ISBN 0-8493-5747-0

A Library of Congress record exists under LC control number: 79017539

Publisher's Note
The publisher has gone to great lengths to ensure the quality of this reprint but points out that some imperfections in the original copies may be apparent.

Disclaimer
The publisher has made every effort to trace copyright holders and welcomes correspondence from those they have been unable to contact.

ISBN 13: 978-1-315-89088-3 (hbk)
ISBN 13: 978-1-351-06998-4 (ebk)

Visit the Taylor & Francis Web site at http://www.taylorandfrancis.com and the
CRC Press Web site at http://www.crcpress.com

INTRODUCTION

Since its first description as an analytical technique almost twenty years ago, radioimmunoassay (RIA) has become increasingly popular for a variety of diagnostically important constituents. A comprehensive discussion of the principles of RIA is beyond the scope of this text since this has been the subject of a number of reviews and articles. The text edited by Odell and Daughaday[1] serves as an excellent source book for the basic principles of RIA. The review by Walker[2] and articles by Zettner and Duly[3,4] give detailed discussions on theoretical considerations and principles of method optimization.

The popularity of this technique arises from two fundamental properties inherent in the approach. First, the technique is exceedingly specific, which makes possible the analysis of single components present in complex matrices (e.g., serum) even when components have very similar structures. A classic example is the anti-arrythmic drug, digoxin, which differs from another therapeutic analogue, digitoxin, by a single oxygen atom at position 12 of the steroid nucleus, yet cross-reactivity between the two species is often less than 1.0% using suitable antibodies. Although radioimmunoassay is a specific term which implies the use of antibodies as binders in the analytical system, a variety of other binders have been used, with high specificity, in competitive binding assays. Proteins normally present in plasma, such as cortisol-binding globulin, and thyroxine-binding globulin, as well as normal tissue receptor proteins such as intrinsic factor and uterine receptors have been used in analytical schemes for the analysis of cortisol, thyroxine, vitamin B_{12}, and estrogens, respectively. Assays that use these types of binders have been commonly called competitive protein binding assays. In an effort to standardize nomenclature, the term saturation analysis has been proposed to encompass all approaches in which naturally occurring or immunologically derived binders are used in an analytical scheme.

The second inherent property of these kinds of analyses is the excellent analytical sensitivity. Since the binders in these assays may have association constants for their ligands as large as 1×10^{10} ℓ/mol, the concentrations which can be detected using these binders can be in the picomolar range. The only other approach for the analysis of compounds in this concentration range is bioassay which often suffers from poor reproducibility, long analysis times, and excessive cost. In the past, the most common means for distinguishing between bound and free ligands in radioimmunoassay and competitive protein-binding techniques has been to use radioactive analogues of the ligands and to separate free and bound fractions by physical means. Recently a variety of other labels have been used (e.g., fluorescence, electron spin resonance, and enzymatic), which have the advantage of not requiring physical separation of free and bound fractions. In these instances the fluorescence intensity or polarization, the resonance signal, or the enzyme activity in the respective cases is different when the labeled material is free in solution than when it is bound to antibody or binder. These methods, although they have the similar theoretical specificity, cannot achieve the theoretical sensitivity which can be accomplished using radioactive labels.

Due to the unique specificity and sensitivity of these techniques, the analytical capability for many diagnostically important constituents in body fluids was realized. Widespread utilization of many of these methods however, was hampered by the availability of suitable antibodies and radioactively labeled analogues. Once manufacturers were able to supply these components along with other necessary constituents in "kit" form, the analytical capability was extended to almost any laboratory with a radioactivity counter and without the need for personnel skilled in immunologic methods. The overall approach, however, remained labor intensive, with many possible sources for error. In order to alleviate some of these problems, several of the analytical steps including sample and reagent processing, radioactivity counting and post analysis data reduction became more automated. High-speed samplers, dispensers, and counters, in addition to on-line and off-line data reduction, helped speed the analysis and eliminate some sources of error. For the most part, however, the separation of free and bound ligand fractions remained a manual operation. The difficulty of automating this phase of the analysis, was perhaps the major drawback to the development of totally automated systems for radioimmunoassay.

Recently, however, automated approaches have been devised for separating free and bound ligands along with the other steps in radioimmunoassay. The various steps which all of these radioassays have in common and must be incorporated into any automated system are the following:

- Introduction of the sample (e.g., serum or other biological fluid, calibrating standards, and controls)
- Addition of fixed amount of specific antibody or binding protein
- Addition of fixed amount of radioactively labeled ligand (analogue of molecule to be measured)
- Incubation of the mixture during which time labeled and unlabeled ligand compete for binding sites on the antibody or binder
- Separation of bound (labeled and unlabeled) ligand from that which is free in solution
- Measurement of radioactivity in the separated fractions in a gamma or beta counter
- Calculation of the concentration of analyte in the unknown sample from a standard curve prepared with known amounts of unlabeled ligand (analyte)

The various instruments described in this text all provide an integrated and automated approach for performing these various steps. Although the various systems differ significantly in design, they all perform these basic manuevers, for the most part, without any operator intervention. It is hoped that the following descriptions of the various system capabilities will not only provide useful information for present and potential users of the systems, but also stimulate additional development by manufacturers which will bring the full potential of radioimmunoassay even closer to a reality.

REFERENCES

1. **Odell, W. D. and Daughaday, W. H., Eds.,** *Principles of Competitive Protein-Binding Assays,* J. B. Lippincott, Philadelphia, Pa. 1971.
2. **Walker, W. H. C.,** *Clin. Chem.,* 19, 146, 1977.
3. **Zettner, A.,** *Clin. Chem.,* 19, 699, 1973.
4. **Zettner, A. and Duly, P. E.,** *Clin. Chem.,* 20, 5, 1974.

THE EDITOR

Lemuel J. Bowie, Ph.D., is Director of the Division of Clinical Biochemistry in the Department of Pathology and Laboratory Medicine at Evanston Hospital, Evanston, Illinois. In addition, he holds an appointment as Assistant Professor of Clinical Pathology at the Northwestern University Medical School, Chicago, Illinois.

Dr. Bowie received his Bachelor of Sciences degree in medical technology from Xavier University of New Orleans, Louisiana in 1966. He worked for one year as a supervisory medical technologist at Providence Hospital in Washington, D.C. prior to matriculating as a graduate student at Johns Hopkins University, Baltimore, Maryland, in 1967. He obtained his Ph.D. in biochemistry in 1972. After a year of post-doctoral training in Clinical Chemistry at the Veterans Administration Hospital, San Diego, California, he stayed on to become Assistant Chief of the Clinical Chemistry Section, Laboratory Service and Assistant adjunct Professor of Chemistry and Pathology, University of California, San Diego.

Dr. Bowie is a diplomate of the American Board of Clinical Chemistry and a member of the American Association for Clinical Chemistry and the National Academy of Clinical Biochemists. He is also a member of the American Chemical Society, the American Association for the Advancement of Science, the Academy of Clinical Laboratory Physicians and Scientists, the International Society for Clinical Enzymology, and the Subcommittee on Enzyme Assay Conditions of the National Committee for Clinical Laboratory Standards. He was the recipient of the 1978 Presidential Citation Award from the American Association for Clinical Chemistry.

Dr. Bowie has had a longstanding interest in automated instrumentation and has coauthored a series of featured reviews on selected categories of clinical laboratory instrumentation in *Analytical Chemistry.*

Dr. Bowie became Director of Clinical Biochemistry at Evanston Hospital July 1979.

CONTRIBUTORS

Lemuel J. Bowie, Ph.D.
Director
Division of Clinical Biochemistry
Evanston Hospital
Evanston, Illinois
Assistant Professor of Clinical Pathology
Northwestern University Medical School
Chicago, Illinois

Gary Brooker, Ph.D.
Professor of Pharmacology
University of Virginia School of Medicine
Charlottesville, Virginia

Edward Cohen, M.BA.
Director of Immunology Marketing
Technicon Instruments Corporation
Tarrytown, New York

William A. Eisenhardt, Ph.D.
Medical Products Division
Union Carbide Corporation
Rye, New York

Gerhard Ertingshausen, Ph.D.
Manager
Clinical Chemistry and Immunoassay
Medical Products Division
Union Carbide Corporation
Rye, New York

Robert Henry
Biosystemès, S.A.
Union Carbide
Montpelliér
France

Sam W. L. Hitch, B.S., M.BA.
Instrumentation Development Manager
Squibb Institute for Medical Research
Princeton, New Jersey

LaVell R. Johnson, Ph.D.
Director of Research
Becton Dickinson Immunodiagnostics
Salt Lake City, Utah

W. Geoffrey Lawn, Ph.D.
Technical Coordinator
Micromedic Systems
Horsham, Pennsylvania

Donald H. Lo,
Medical Products Division
Union Carbide Corporation
Rye, New York

Bridgette Mériadec,
Biosystèmes, S.A.
Union Carbide
Montpelliér
France

D. R. K. Murty, Ph.D.
Assistant Director, Clinical Assay Development
Squibb Institute for Medical Research
Princeton, New Jersey

Steven L. Ossad, M.A., M.B.A.
Product Manager
Immunodiagnostics
Technicon Instruments Corporation
Tarrytown, New York

William Piknosh, B.A.
Applications Chemist
Clinical Laboratory
Picker Corporation
Northford, Connecticut

Frank Savino, B.S.E.E., M.E.E.,
Systems Engineer
Technicon Instruments Corporation
Tarrytown, New York

Stephen I. Shapiro,
Medical Products Division
Union Carbide Corporation
Rye, New York

Mary Lu Simons, M.A.
Product Manager
Clinical Laboratory Department
Picker Corporation
Northford, Connecticut

TABLE OF CONTENTS

Chapter 1

THE UNION CARBIDE CENTRIA® SYSTEM

G. Ertingshausen, S. I. Shapiro, W. A. Eisenhardt, D. H. Lo, B. Mériadec, and R. Henry

TABLE OF CONTENTS

I. INTRODUCTION

During the first Annual Symposium on "High Resolution Analyses and Advanced Concepts for the Clinical Laboratory" held in Oak Ridge, Tennessee in 1969, Norman G. Anderson reported on a new approach to high speed automated spectrophotometric analysis.[1] His Gemsaec or Fast Analyzer concept uses centrifugal force to mix rapidly multiple samples and reagents during transfer into an array of cuvettes which are located at the circumference of a rotor. The cuvettes spin past the stationary light beam of a photometer which continuously measures the absorbance, displays it on an oscilloscope, and processes the data according to programs provided by a mini-computer.

This unique concept initiates multiple reactions simultaneously, processes a batch of standards and samples under identical reaction conditions of time and temperature, and provides an excellent tool for absorbance measurements under nonequilibrium conditions since the reactions are continuously available for monitoring.

When Union Carbide took on the commercialization of this idea another inherent unique feature of the concept was exploited: the ability to take absorbance measurements as early as 2 sec after mixing of the reactants. The absorbance values taken at this time are in many cases identical to the sum of the blank absorbances of serum sample, reagent, and individual cuvette. They could be automatically subtracted from the final absorbance after completion of a reaction and no further blank was required.

In pursuit of this autoblanking capability, and to ensure that samples are completely transferred and mixed with reagent, Anderson's original rotor (Figure 1) was changed to a configuration (Figure 2) in which the reagent washes the sample out of its cavity into the cuvette. The new disc no longer required a high initial speed for liquid transfer and a lower one for subsequent, delayed absorbance readings.

It occurred to us as early as 1971 that some of the characteristics which made the centrifugal analyzer concept attractive for batch chemical analysis could be applied to the mechanization of radioimmunoassay. Simultaneous initiation of all reactions is desirable because it avoids "first-to-last-tube-differences" in those cases where a sequential pipetting and mixing process is followed by simultaneous centrifugation of precipitates of charcoal or double antibody complex. Also, using centrifugal force to transfer and mix multiple solutions provides a mechanically elegant and reliable system.

A capability to stop all reactions at the same time after simultaneous combination of all reactants is also attractive for RIA so that an assay may be performed under nonequilibrium conditions with the concomitant savings in assaying time. A simple way to interrupt multiple incubations by a quick separation process for bound and free label was needed. This had to be achieved by bringing multiple incubation mixtures in contact with a separation medium quickly and by subsequently separating bound and free fractions simultaneously for all incubates.

Combining all these requirements and taking a close look at the modified Centrifichem®* transfer disc suggested the following thoughts for an automated RIA system:

1. Transfer antibody reagent at low rotational speed from inner cavity to outer cavity filled with sample and radioactive ligand.
2. Hold mixture in outer cavity during incubation at lower speed.
3. Accelerate and move liquids into transfer channel which holds a separation medium.
4. Rinse disc by dynamically introducing eluant and collect effluents from separator medium.
5. Count effluent containing "free" or "bound" ligand on separate counter module.

These thought processes were translated into an experimental prototype[2] which later became the Incubator/Separator of the Union Carbide Centria® System (Figure 3).

* ® Trademark of Union Carbide Corporation.

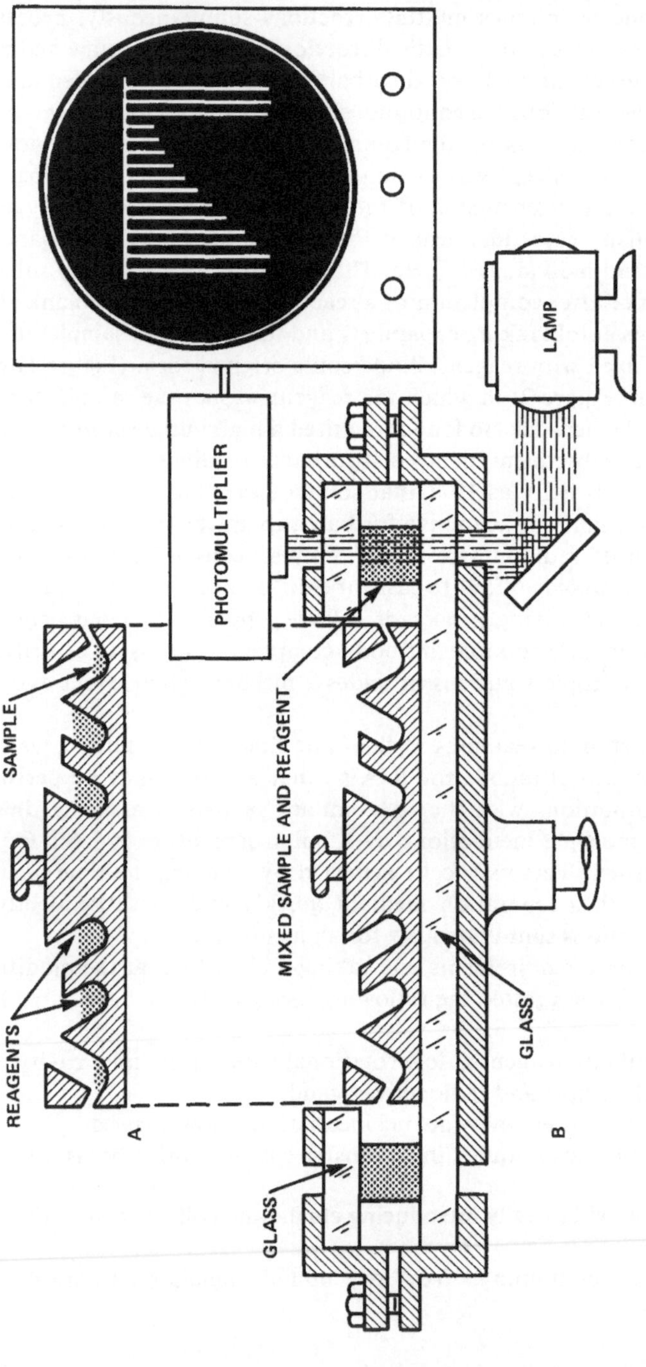

FIGURE 1. Principles of operation of the GEMSAEC® Fast Analyzer. (From Anderson, N. G., *Science*, 166, 317, 1969. With permission.)

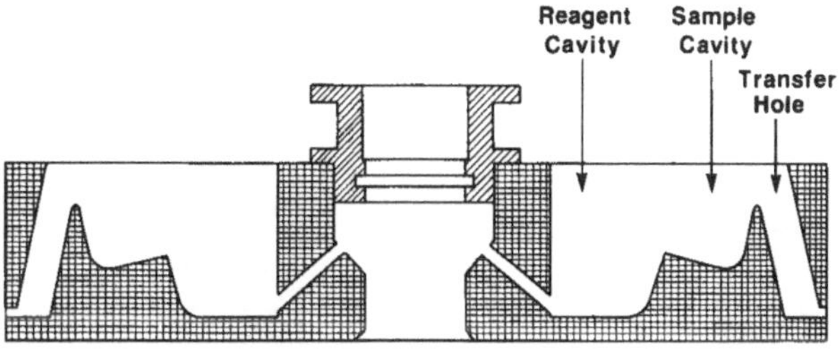

FIGURE 2. Reagent and sample holder for Centrifichem®. (From Fabiny, D. L. and Ertingshausen, G., *Clin. Chem.*, 17, 696, 1971. With permission.)

II. THE PRINCIPLE OF CENTRIA® SYSTEM'S AUTOMATED INCUBATION AND SEPARATION

The Centria® Incubator/Separator is the heart of the system, where incubation and separation of the reactants are performed (Figure 4). It consists of a variable-speed turntable, which is first loaded with a plastic ring holding 36 small ready-to-use separating columns packed with separation medium. They are positioned in test tubes, which in turn are suspended by their upper rims in ball seats in the plastic ring. The filled transfer disc is placed on the turntable and automatically keyed such that each pair of cavities is aligned with the opening of a separating column.

Figure 5 shows a single sample position. When rotation of the turntable is started, antiserum flows from the inner into the outer cavity, where all reactants are instantaneously mixed and held for a preset interval. The steep angle and overall shape of the outer cavity prevent any overflow at the first speed. After the incubation time has elapsed, the instrument automatically accelerates the turntable rotation to a second, higher speed, causing the incubated mixture to be transferred onto the columns, where it embeds in the separating medium. After embedding has taken place, a pump is started, which dispenses a selected amount of liquid onto the spinning transfer disc, where it is divided into equal aliquots by the geometry of the disc. While the eluant is driving the incubated mixtures through the separating medium, the free radioactive tracer is either adsorbed or partitioned out on the column. The antibody-bound tracer in turn moves with the front of the eluant and is collected in the test tubes that support the spinning columns. The elution process takes about 60 sec. After the turntable has come to rest, the transfer disc is discarded and the column ring holding the column/test tube combinations is lifted and transferred to the turntable of the Counter/Computer.

Sephadex® G-25 and Sephadex® DEAE-A25* were successfully used to separate molecules of low molecular weight from their antibody complexes. Sephadex® DEAE-25 combines both gel filtration and ion exchange characteristics and has led to a significant reduction of apparent nonspecific binding values in radioimmunoassays for T_4 and T_3 when compared to G-25. DEAE-50 columns were successfully used to separate large molecules like insulin, growth hormone, and human placental lactogen. This separation principle failed when applied to the separation of LH and FSH from their respective antibody complexes. Furthermore the Sephadex® DEAE-50 columns could not easily be produced and tended to compress when spun in the Centria® Incubator/Separator.

When these limitations became apparent, R. Henry and B. Mériadec developed an

* ® Trademark of Pharmacia Corporation.

FIGURE 3. The Union Carbide Centria® System.

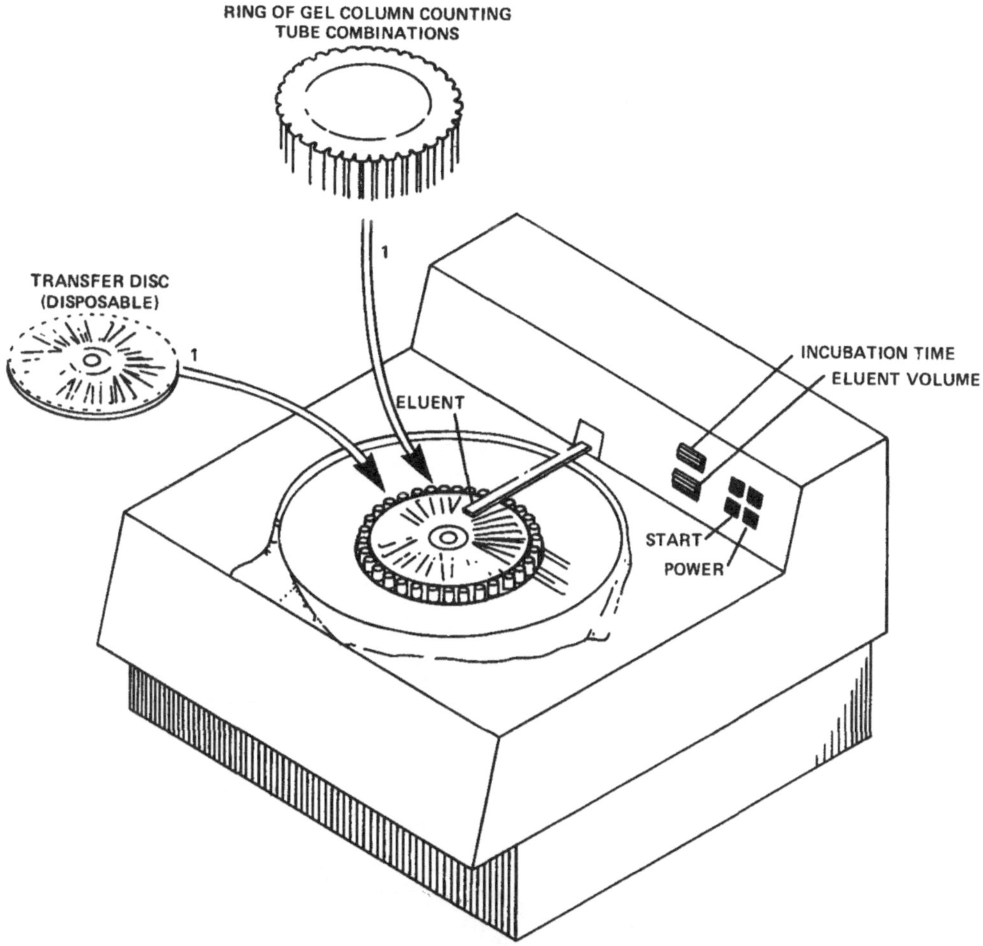

FIGURE 4. The Centria® Incubator/Separator-Prototype. (From Ertingshausen, G., Shapiro, S. I., Green, G., and Zborowski, G., *Clin. Chem.*, 21, 1305, 1975. With permission.)

additional separation procedure based on Sepharose-bound second antibody, which is illustrated in Figure 6. Sepharose-bound anti-IgG antibody is tabletted and placed on top of a hydrophobic disc located in the bottom of a plastic column. The columns are suspended in a special metal ring which enables the operator to remove the columns without lifting the test tubes. This special ring is supported by the Centria® test tube holding ring.

When the incubation of a typical Centria® System assay is terminated the incubates are transferred to the columns as usual. The liquid submerges the anti-IgG tablet which swells instantaneously trapping all liquid in the gel. The rotor returns to the first speed. After a 5 min incubation, the gel is rinsed with eluant buffer. This occurs at the second incubator/separator speed. After about 2 m*l* of eluant are distributed to each column sufficient pressure builds up to break the hydrophobic barrier of the disc at the bottom of the column and eluant is collected in the supporting test tubes. The rinsing process is repeated once. When this kind of separation is applied the columns which retain the bound phase are counted instead of the test tubes.

III. SYSTEM DESCRIPTION

The Centria® System was the first analytical device to provide an integrated ap-

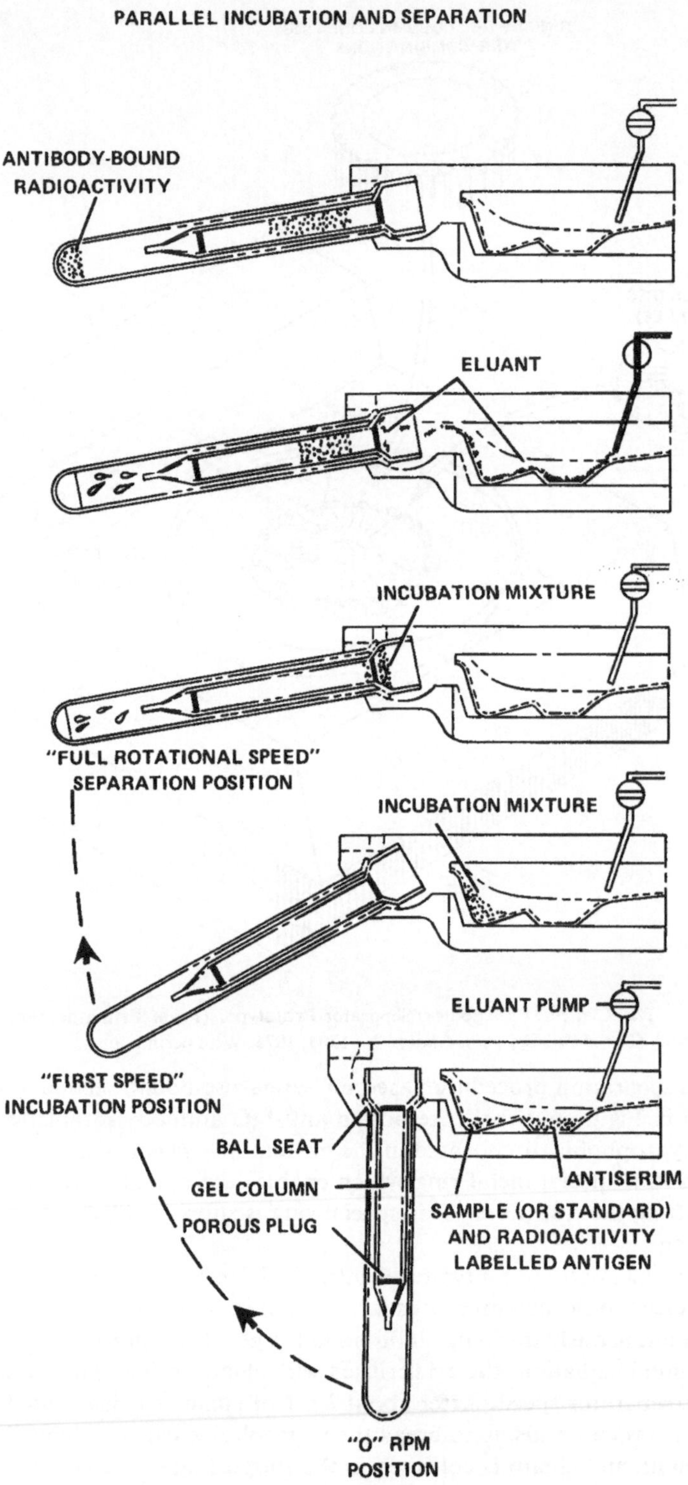

FIGURE 5. Side view of single sample position in the Centria® Incubator/Separator performing column separation. (From Ertingshausen, G., Shapiro, S. I., Green, G., and Zborowski, G., *Clin. Chem.*, 21, 1305, 1975. With permission.)

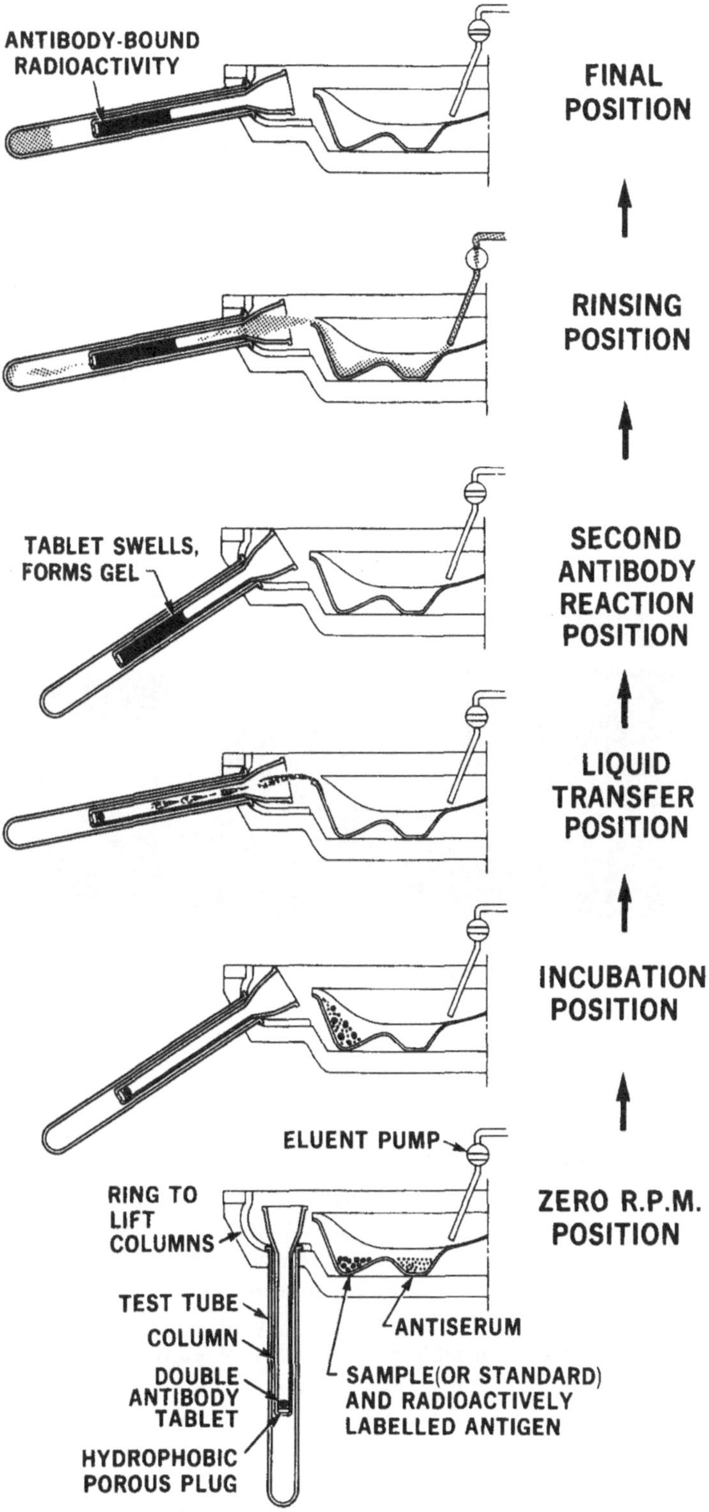

FIGURE 6. Side view of a single sample position in the Centria® Incubator/Separator performing a double antibody tablet separation.

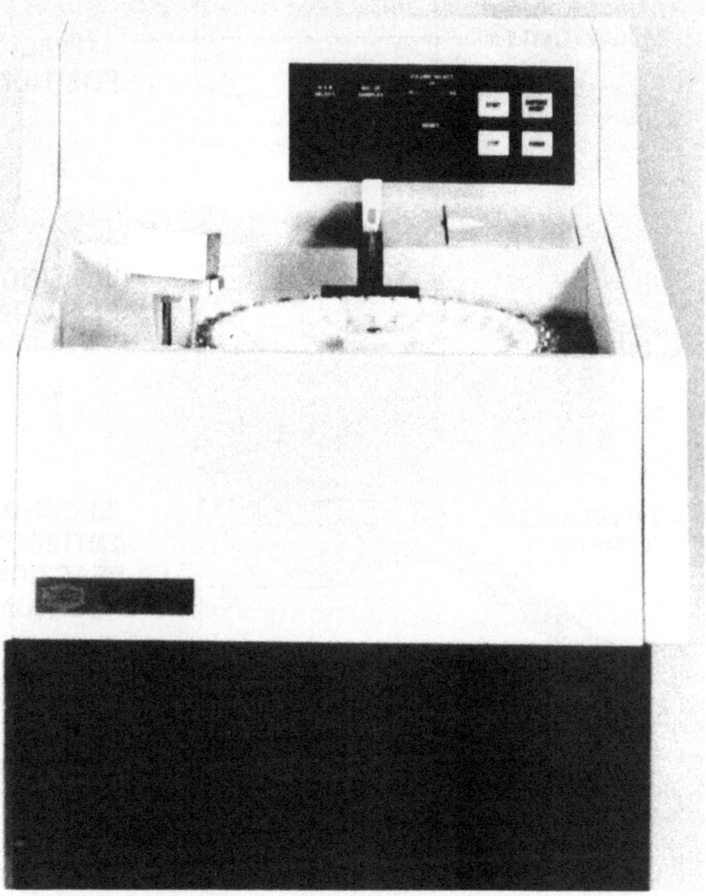

FIGURE 7. The Centria® Pipettor.

proach to performing radioimmunoassays. It automates all critical manual steps associated with manual RIA procedures and combines all these operations in a single instrument.

A. The Pipettor and Transfer Disc

The Centria® Pipettor (Figure 7) is designed around a turntable which is first loaded with a metal ring holding at its circumference a maximum of 36 sample cups. The ring fits onto the Pipettor turntable in only one properly indexed position which is found by matching the hole in its base to the locating pin on the surface of the turntable. The same locating pin is also used to properly position the transfer disc which fits inside the sample ring. The disposable transfer disc molded from polyethylene, consists of 36 radially aligned pairs of cavities (Figure 8).

Standards and samples are automatically withdrawn from the sample cups holding about 250 $\mu\ell$ of liquid and are then loaded into the outer cavities of the disc. The sample volume is either 15, 35, or 50 $\mu\ell$, depending on the assay. In each case the sample is diluted to 100 $\mu\ell$ to complete transfer of the measured volume and to minimize carryover. Before the next sample is aspirated, the outside of the sample tip is rinsed in a reservoir with distilled water which is continually replenished from a large bottle. While sample or standard are being loaded, 200 $\mu\ell$ antiserum and 50 $\mu\ell$ radio-

FIGURE 8. The Centria® Transfer Disc.

active competing antigen are automatically withdrawn from their respective reagent bottles and dispensed into the radially aligned cavities — antiserum into the inner cavity and radioactive liquid into the outer cavity. This is achieved by a second pumping mechanism. Provisions are also made for substituting the antiserum with buffer in those positions for which a nonspecific binding count has to be established. The bottle holding tray simply rotates at the appropriate time by 180° around an axis located in the center of the bottle containing radioactive tracer, which thus remains in its original position. Because the system uses air displacement pumps, purging and priming of the measuring systems is not required and hence little reagent is wasted. Antigen or radioactive tracer are loaded only into positions receiving sample, standard, or control serum.

1. The Pipettor Control Panel

Power — A pushbutton control which is lit when on. Power is delivered when this switch is activated.

Reset — A button which must be pushed prior to each pipetting cycle. It resets the logic of the pipettor so that all positions are pipetted according to setting on the control panel.

Start — The pipettor arms will begin dispensing reagent and sample when this button is pushed.

Stop — At any time after the cycle is initiated, pipettor action can be halted by pressing the STOP button. The pipettor will stop all activity when it finishes its current position.

Dispense inhibit — When this button is activated during a cycle, pipettor will continue its movement, but nothing will be picked up or dispensed. This feature can be used to check out the mechanical function of the pipettor at the start of the day, or to stop dispensing reagent during a pipetting cycle in case of operator error.

NSB select — These two thumb dial controls are used to program the pipettor to pick up buffer instead of antibody. Setting 02 programs the instrument to pipet positions 1 and 2 as NSB positions. This is accomplished by the rotation of the reagent turntable 180° to position the buffer vial under the reagent tip.

Number of samples — This pair of thumb dial switches is used to program the number of positions on the disc that are to be pipetted. The number 36 is dialed in for a full disc. If only a partial disc is desired, that number is dialed in and positions after this point will not receive any sample or reagent.

Sample volumes — This 3-position dial is set at 15, 35, or 50 $\mu \ell$ depending on the specific assay. The programmed volume is automatically picked up from each sample cup and delivered to the transfer disc, followed by deionized water diluent. Total volume delivered is 100 $\mu \ell$.

B. The Incubator/Separator

This module is the key to the system where incubation and separation are performed simultaneously (Figure 4). Transfer discs filled with reactants are processed here according to the principle described above. The turntable rotates at 95 r/min at its first and at 210 r/min at its second speed. For assays requiring long reaction times, initial mixing is performed in this module. Discs are then removed and placed in a separate incubator. They are placed back on the turntable after the incubation is over and then processed through the separation cycles. A peristaltic pump is used to pump eluant buffer on the rotating transfer disc.

1. The Incubator/Separator Control Panel

The incubator/separator control panel consists of four pushbutton switches and three thumb wheel dials.

Power — Activate: light will come on indicating power is being delivered to the module.

Eluant on — Press and hold down button to prime eluant pump line.

Eluant Time — Controls the length of time the eluant pump will operate. The dial is calibrated in tenths of minutes. Dial the appropriate time setting to deliver the desired volume of eluant onto the disc (e.g., 11 directs the pump to operate for 1.1 min).

Incubation Time — Directs the centrifuge to rotate at 95 r/min for a specific length of time. The two digit dial is calibrated in minutes. Each RIA will have a unique incubation time. Refer to the assay method sheet for the correct setting on this dial (e.g. 15 directs the centrifuge to rotate at 95 r/min for 15 min; 30 directs the centrifuge to rotate at 95 r/min for 30 min).

Second incubation (for tablet separation only) — Directs the centrifuge to rotate at 95 r/min for a specific length of time for a second time after transfer of incubate onto column. During this time the separation medium may interact with the incubate. If "00" is dialed in for separation time rotor does not return to 95 r/min but stays at 210 r/min until separation is completed.

Start — Press to initiate incubation and separation sequence. Cycle will progress automatically from this point unless interrupted manually.

Stop — Activate to interrupt the incubation and separation sequence. The centrifuge will stop spinning immediately.

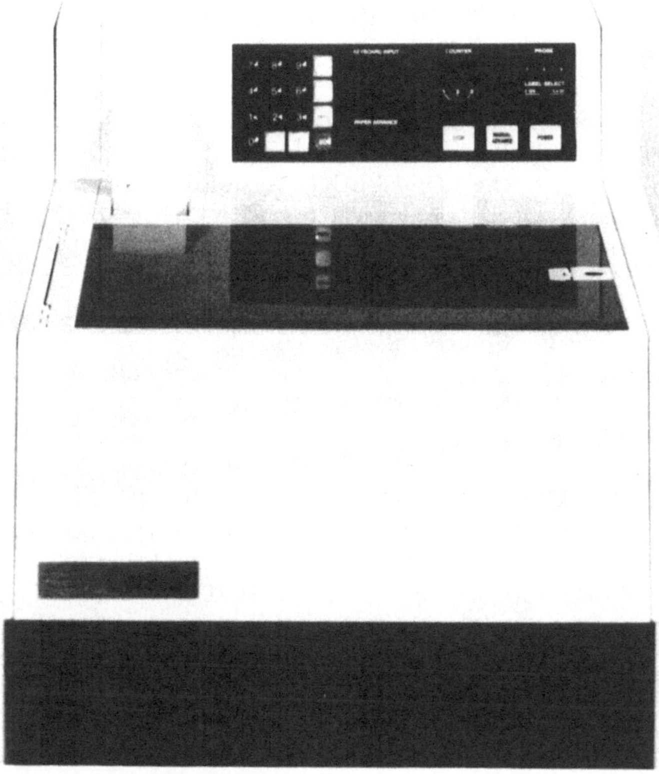

FIGURE 9. The Centria® Counter/Computer.

C. The Counter/Computer

The counter — The third module of the system consists of a three-channel gamma counter/computer combination whose function is to count the radioactivity in the test tubes or separation columns and to calculate concentrations based on counts from standards run simultaneously with the clinical samples (Figure 9). The rings holding either column/test tube combinations or columns with immobilized second antibody alone are placed on the counter turntable, which, at the beginning of a run, is in its elevated position. Keying the column ring to the turntable assures the proper assignment of counting positions of all standards and samples. Test codes, counting time, standard values, and method of data reduction are then entered into the microprocessor through an interactive keyboard. When the start button is pushed, the turntable indexes and descends to the lower position, and three test tubes or columns at a time are guided mechanically into three counting wells, which are fixed 120° apart. The three counting systems are 1.75 × 2 in. sodium iodide crystal/photomultiplier tube combinations. When the first counting interval has elapsed, the turntable lifts up, advances, and lowers the next set of three tubes into the probe. This change-over takes 8 sec. At the end of the counting cycle, the counts are printed out or processed on-line, according to the initially chosen data reduction method which converts them into concentration units. The counter is designed to count iodine-125 and cobalt-57.

Data analysis — There are four data transformations available for calculations on the Centria® Counter/Computer. These are

1. LOGIT-LOG

2. HYPERBOLA
3. ARC-SIN
4. PROBIT

It is also possible to have unreduced data printed out as another "option". When running a standard assay, all parameters which govern counting, collation, and reduction are preprogrammed. The LOGIT-LOG transformation is used for all standard runs because of better correlation and widespread use. If the LOGIT-LOG transformation is not desirable, the HYPERBOLA, ARC-SIN, or PROBIT transforms may be chosen from the optional group of parameters.

1. The Counter/Computer Control Panel

The control panel consists of six switches and a 16-button keyboard. The function and operation of the switches are as follows:

Power — This is a push-button switch that controls power to all functions of the counter/computer (C/C) except the photomultiplier tubes (PMTs.) Power is delivered to the PMT's whenever the module is plugged in and the circuit breaker is in ON position. A light inside the switch indicates that power is being delivered to the C/C. Power must be on to perform any operation on the C/C.

Isotope select — A two-position switch that regulates the pulse discriminators for maximum counting efficiency of the radioisotope being used. ^{125}I or ^{57}Co is selected according to the specific assay being counted.

Manual advance — Regulates the turntable independent of computer direction, however, the computer is aware of manual advance signal and will operate accordingly. Directly controls the motor of the turntable.

Stop — A pushbutton used to interrupt any sequence of the C/C when needed instead of turning off power. When activated, turntable motor will stop — computer is aware of stop signal and usually requires additional response from the operator to resume processing.

Paper advance — Indexes the printer paper under the computer's control without printing any data. A contact sensor will detect if the printer is out of paper and stop the printer until more paper is inserted.

Keyboard — Means of conversing with the computer. Numerals 0 to 9, plus decimal point, YES, NO, ENTER, and CLEAR buttons are found on the keyboard. Three LED displays inform the operator of the current activity of the counter.

1. Keyboard input — displays data entered on keyboard by operator
2. Counter — displays numerical counts for only the most recently counted tube in a specific probe. The switch can be set at 1, 2, or 3 at any time to monitor the tubes without interrupting the counting sequence
3. Probe indicator lights — frequency of flashing indicates rate of radioactivity being detected by each probe; each ON or OFF change of the light indicates 64 counts detected by the probe

Probe calibration potentiometer — Available to the operator to properly calibrate the three probes for matching and peak efficiency. The three potentiometers are adjusted with a radioactive source in the corresponding well. The RATE of activity is noted on the meter dial and the meter dial brought to peak deflection using the corresponding calibration potentiometer. This assures linearization and scaling for the PMT's. The calibration potentiometers affect the relationship that exists between fixed window voltages and corresponding energy values; i.e., 0.85 V on the lower discriminator corresponds to 15 KeV in energy.

D. Data Presentation

After selecting the appropriate data handling parameters on the keyboard of Counter/Computer, the counting process starts. Once it is terminated the data are presented according to the format now described in detail.

1. The Printout

ENTER COUNT TIME (SECONDS)	20 to 600
ENTER UPPER LIM	Anything above this limit will be flagged HHH (high)
ENTER LOWER LIM	Anything below this will be flagged LLL (low)
NUMBER OF TUBES	Must be *at least* the number of standard tubes that have been requested.
START COUNTING?	C/C asks
YES	If you wish to begin, press *yes.* Turntable will move to "home" position and counting will begin. After 13.5 min, results will be printed (for 60 sec counts).

The Centria® Counter/Computer will print out the following information when all tubes have been counted.

T_4 RIA ←	Assay being performed
RUN = 002←	Run number for the day
May 17, 1976←	Date entered after power on
LOGIT-LOG	Transform selected
NG/ML	Concentration units
CORR 0.99822	Correlation coefficient "r"
SLOPE −1.05076	Slope of standard curve
INCPT 4.30350	Y intercept of standard curve
50% BO 60.07 NG ←	Concentration that will cause a 50% dis-
O1 NSB CTS 1246←	placement of the label from the antibody
02 NSB CTS 1398	Counts in nonspecific binding tubes
STANDARDS	
STD # 1 = 0.00 ←	Assigned std. concentration
03 0 CTS 07247←	Counts in the first replicate of 0 std.
04 0 CTS 07206←	Counts in the second replicate of 0 std.
STD # 2 = 30.00←	Assigned conc. of std. #2
05 30.87←	Conc. of std. #2 from actual data point
06 31.23←	Conc. of replicate std. #2 from actual data point
31.05/ 0.6% ←	Av of replicates/% variance of replicates from the mean
STD. #3 = 60.00	
07 58.18	
08 58.18	
58.18/ 0.0%	
STD. #4 = 120.0	
09 110.2	

Printout	Explanation
10 120.4	
115.3/ 4.4%	
STD. #5 = 240.0	
11 258.5	
12 240.3	
249.4/ 3.7%	
UNKNOWNS	
⋮	
13 31.77	
14 35.45	
33.61/ 5.5%	
15 63.43 ←	Conc. in patient sample
16 63.72 ←	Replicate patient sample
63.57/ 0.2% ←	Av of replicates/% variance of replicates from the mean
17 117.2	
18 121.0	
119.1/ 1.6%	
19 258.0	
20 257.1	
257.1/ 0.2%	
21 47.49	
22 46.69	
47.09/ 0.9%	
23 140.6	
24 137.3	
138.9/ 1.2%	
25 143.4	
26 153.1	
148.2/ 3.3%	
27 49.13	
28 52.03	
50.58/ 2.9%	
INTRARUN PRECISION	
2.0% ←	Av of all the % variance of replicates for the entire disc
FLAGS	
?? = CONC. EXCEEDS STD. CURVE	
<< — PT. POSSIBLY AN OUTLIER ←	Standard value deviated by 10% or more from the theoretical concentration
HH = ABNORMAL HI CONC. } ←	Concentration is outside the normal limits determined by operator
LL = ABNORMAL LO CONC. }	
TECH ←	Space for Tech. Signature
Raw Counts?	After results are printed, raw counts can be printed out.

ANOTHER COPY?	C/C asks	
NO	Enter yes or no	
RECOMPUTE?	C/C asks	
YES	Enter yes or no	
RAW COUNTS?	C/C asks	
T_4RIA	Enter yes or no	
RUN = 0.02		
MAY 17, 1976	The instrument background has been subtracted from these counts	
COUNTS %		
60 SEC B/BO		
01 001246	Time each tube is counted	Ratio of bound to bound of "0" standard (expressed in %)
02 001398		
03 007247		
04 007206		
05 005267	66.8%	
06 005251	66.5%	
07 004324	50.8%	
08 004324	50.8%	
09 003364	34.6%	
10 003242	32.5%	
11 002370	17.7%	
12 002438	18.9%	
13 005227	66.1%	
14 005072	63.5%	
15 004190	48.6%	
16 004183	48.5%	
17 003278	33.1%	
18 003235	32.4%	
19 002372	17.8%	
20 002375	17.8%	
21 004637	56.1%	
22 004663	56.6%	

Recalculation of the Standard Curve After the printout, the operator has the option to:

1. Change the transform
2. Delete one or more of the values used in the calculation of the standard curve.

The procedure is as follows:

T_4 RIA
RUN = 008
MAY 10, 1976
LOGIT-LOG
μg/dl
CORR 0.99767
SLOPE −1.05201
INCP 1.96656

50% BO 6.48 μg	
BO/T 62.6%	
01 NSB CTS 1395	
02 NSB CTS 1408	
03 TOTALS 012591	
04 TOTALS 012744	
STANDARDS	
STD. #1 = 0.00	
05 0 CTS 09440	
06 0 CTS 09223	
STD. #2 = 3.00	
07 2.94	
08 2.98/1.4%	
STD. #3 = 6.00	
09 6.02	
10 6.29	
6.16/2.1%	
STD. #4 = 12.00	
11 11.24	
12 12.07	
11.65/3.6%	
STD. #5 = 24.00	Printout indicates that tube #14 deviates by more than 10% from the theoretical concentration
13 22.24	
14 26.49 <<<	
24.37/8.7%	
ANOTHER COPY	C/C asks
NO	Enter yes or no
RECOMPUTE	Do you wish to recompute the standard curve?
YES	
CHANGE TRANSFORM?	At this point, the operator can recalculate the standard curve with another equation if desired Enter yes or no
COUNTS?	Enter yes or no
SELECT TUBES TO BE DELETED FROM COMPUTATION	
ENTER TUBE #	
14	Enter tube to be deleted
DELETED	#14 is deleted
0	Entering "0" indicates that you do not wish to delete further
T_4RIA	
RUN = 0.08	Computer now prints out data using the new calculation
MAY 10, 1976	
LOGIT-LOG	
μg/dℓ	
CORR 0.99888	
SLOPE −1.01738	
INCPT 1.91251	

50% BO 6.55 µg	
BO/T 62.2%	
01 NSB CTS 1395	
02 NSB CTS 1408	
03 TOTALS 012591	
04 TOTALS 012744	
STANDARDS	
STD. #1 = 0.00	A maximum of 4 data points may be deleted
05 0 CTS 09440	
06 0 CTS 09223	*Note:* One set of duplicates and two additional single data points
STD. #2 = 3.00	*Note:* Only one of the "0" standards may be deleted
07 2.98	
08 2.97	
2.93/ 1.4%	
STD. #3 = 6.00	
09 6.07	
10 6.35	
6.21/ 2.2%	
STD. #4 = 12.00	
11 11.57	
12 12.46	
12.02/ 3.7%	
STD. #5 = 24.00	
13 23.44	
14 DELETED	Tube #14 has been deleted
15	All samples are recomputed using
16	the new standard curve
Retaining the Standard Curve	
ANOTHER COPY?	After results are printed out, C/C asks
NO	Enter yes or no
MORE TUBES?	Enter yes or no. If yes, the counter/computer will retain the standard curve and calculate more unknowns.
YES	
SAVE PARAMETERS?	Enter yes or no. If yes, C/C will retain the same parameters that were programmed but will not save the standard curve. The next ring of tubes must contain a new standard curve.

IV. THE CENTRIA® SYSTEM METHODS

A. Procedures for Small Molecules and Insulin

The system's principle was conceived with the thought in mind that any liquid phase radioimmunoassay should easily be adaptable to the instrument. For this reason Sephadex® G-25 was chosen as the initial separation medium. It is the best characterized and most reproducible separation material suited for a universal disposable column for the separation of small molecules. The separation is based on gel filtration but on occasion the gel also adsorbs highly polar molecules. The gel column, when properly packed, does not compress in the centrifugal field employed. A separation cycle was designed in which the incubate was first allowed to embed completely in the column

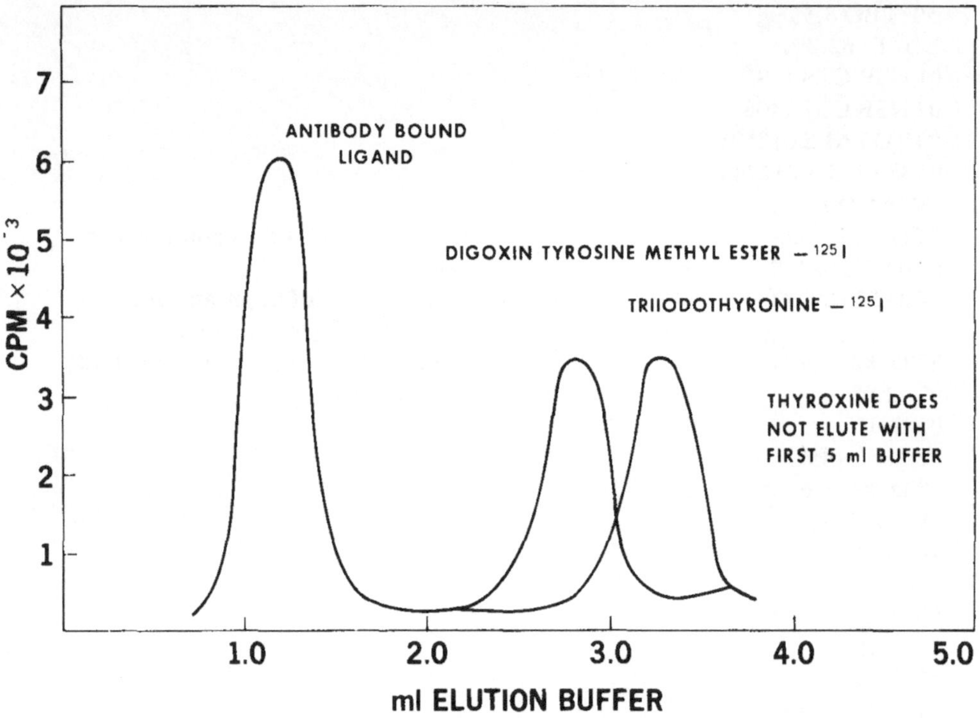

FIGURE 10. Separation of free thyroxine, triidothyronine, and digoxin ligands from their protein complexes.

after being spun off the transfer disc. Subsequently, an accurately timed and, therefore, accurately measured amount of elution buffer was added dynamically to the spinning disc. All macromolecules including the antibody bound radiotracer were thus eluted while all small molecules below MW < 1000 were retained on the column. Initial work showed that this principle could immediately be applied to the radioimmunoassays for digoxin, thyroxine, and triiodothyronine. It was also immediately applicable to the T_3 uptake procedure and a number of nonimmunological radioligand assays.

A typical elution pattern for the radioactive ligands of digoxin and triiodothyronine is shown in Figure 10. These curves are generated by adding 350 $\mu\ell$ of the respective incubated radioimmunoassay mixtures to the columns, eluting successively with 0.2 mℓ increments of buffer, and counting the radioactivity in the fractions collected from the columns after spinning them at the second speed on the incubator/separator module. Triiodothyronine emerges relatively late at pH 5.2 under the conditions of the T_3-uptake test, but, like thyroxine, is completely retained at pH 8.9. DEAE-A25 retains the free T_4 and T_3 labels.

B. Raising of Antisera Against Digoxin, Thyroxine, and Triiodothyronine

1. Antisera

Anti-digoxin was elicited in rabbits with a conjugate of digoxin/human serum albumin, prepared according to the procedure of Smith et al.[3] Ultraviolet analysis of the conjugate indicated a digoxin/human serum albumin molar ratio of approximately 11. The anti-digoxin (Ra 55) used in these experiments was harvested 6 months after a primary injection of 0.2 mg of conjugate, emulsified in Freund's complete adjuvant; we used the low antigen dose and multidermal injection route of Vaitukaitis et al.[4] During the 6-month immunization schedule, the rabbits received three approximately

Table 1
CROSS-REACTIVITY OF ANTISERUM TO TRIIODOTHYRONINE

Hapten	% Cross-reaction[a,b]	% Cross-reaction[a,c]
Triiodo-L-thyronine	100.0	100.0
Triiodo-D-thyronine	100.0	100.0
L-Thyroxine	0.2	0.2
3,5-Diiodo-L-thyronine	0.9	0.8
Triiodothyropropionic acid	9.8	6.9
Diiodo-L-tyrosine	<<0.1	<<0.1

[a] % Cross-reaction is defined as the reciprocal of that molar ratio (multiplied by 100), of the hapten inhibitor to the unlabeled hapten, required to halve antibody binding of the labeled hapten.

[b] Values for % cross-reaction were determined by using Sephadex® G-25 columns to separate free and antibody-bound hapten.

[c] Values for % cross-reaction were determined by using dextran-coated charcoal in a batch method to separate free and antibody-bound hapten.

From Ertinghausen, G., Shapiro, S. I., Green, G., and Zborowski, G., *Clin. Chem.*, 21, 1305, 1975. With permission.

equally spaced booster injections of 0.1 to 0.5 mg of conjugate emulsified in adjuvant. The harvested anti-digoxin showed a titer (the final dilution of raw antiserum in the incubate that will bind 50% of the radiolabeled hapten) of 110,000. Antibody specificity studies (expressed in terms of percentage cross-reactivity and defined in Table 1) show the antiserum to be highly specific for digoxin. The cross-reactivity, with digitoxin and spironolactone as inhibitors, was 4.3 and < 0.7%, respectively. In addition, three commercially obtained control sera, known to be negative for digoxin, showed insignficant displacement of radiolabeled digoxin in the incubate mixture. The average intrinsic affinity constant of this antiserum, determined by the equilibrium dialysis procedure of Smith et al.[3] was 1.7×10^9 ℓ/mol.

Anti-triiodo-L-thyronine (T_3) was elicited in rabbits with a triiodo-L-thyronine/bovine serum albumin conjugate, synthesized by the procedure of Gharib et al.[5] Ultraviolet analysis of the conjugate showed a T_3-bovine serum albumin molar ratio of 4 to 5. The anti-T_3 (Ra 94-I) used in these experiments was harvested 4 months after a primary injection of 75 μg of conjugate, emulsified with Freund's complete adjuvant; again, we used the low antigen dose and multidermal injection procedure of Vaitukaitis et al.[4] Two boosting injections of 50 μg and 250 μg, emulsified with adjuvant, were given during the 4 month immunization period. The anti-T_3 showed a titer of 1:1750.

The cross reactivity of five inhibitors is shown in Table 1. Similar values for the percentage cross-reaction are seen whether dextran-coated charcoal or the Centria® gel-permeation columns were used in the separation step. The average intrinsic affinity constant for the anti-T_3 was calculated to be 2.4×10^9 ℓ/mol.

Anti-L-thyroxine (T_4) was produced in rabbits with an L-thyroxine/bovine serum albumin conjugate. The method used for conjugation was the same as described for the T_3-conjugate. Analysis of the conjugate showed a T_4/bovine serum albumin molar ratio of 4. The anti-T_4 used in these experiments was harvested 6 months after the primary injection. During the 6 month immunization period, three boosting injections were given subcutaneously and intravenously. The anti-T_4 exhibited a titer of 1:525. Antibody specificity studies indicate an antiserum of low cross-reactivity. The percentage cross-reaction for five inhibitors is seen in Table 2.

Table 2
CROSS-REACTIVITY OF ANTISERUM TO THYROXINE

Hapten	% Cross-reaction*
L-Thyroxine	100.0
Triiodo-L-thyronine	1.0
Triiodo-D-thyronine	1.6
Triiodothyroacetic acid	1.0
Triiodothyropropionic acid	3.0
Diiodo-L-tyrosine	<<0.2

* % Cross-reaction is defined as the reciprocal of that molar ratio (multiplied by 100), of the hapten inhibitor to the unlabeled hapten, required to halve antibody binding of the labeled hapten. Values for % cross-reaction were determined by using Sephadex® G-25 columns to separate free and antibody-bound hapten.

From Ertinghausen, G., Shapiro, S. I., Green, G., and Zborowski, G., *Clin. Chem.*, 21, 1305, 1975. With permission.

Anti-insulin was prepared in young male guinea pigs.[5] As the immunogen we used Iletin® U-40 (a pork insulin), obtained from Eli Lilly & Co., Indianapolis, Ind. 46206. The anti-insulin (GP-68) we used in these experiments was harvested 4 months after the primary injection of 10 USP units of insulin emulsified in Freund's complete adjuvant. During the 4 month immunization schedule, the guinea pigs received three about equally spaced booster injections of 10 units of insulin emulsified in adjuvant. Before each injection, the guinea pigs were given dextrose (Ion-o-trate®; Abbott Laboratories, North Chicago, Ill.) in their drinking water to prevent hypoglycemia. In several cases where the animals appeared to go into shock, they were given an intraperitoneal injection of dextrose (1 mℓ of a 500 g/ℓ dextrose solution). The harvested anti-insulin exhibited a titer of 42,000.

2. Radioligands

Radioiodinated (^{125}I) insulin, high specific activity 3,3′,5′-triiodo-L-thyronine (T_3-^{125}I) and 3,3′,5,5′-tetraiodo-L-thyronine (T_4-^{125}I) for use in the corresponding Centria® System radioimmunoassays, and low specific activity T_3-^{125}I for the System's T_3 Uptake radioassay were prepared and purified by proprietary procedures developed at Union Carbide Corporation, Medical Products Division Laboratories, Tarrytown, New York.[7]

A radioligand for the Centria® digoxin RIA was designed and synthesized using novel, patented[7] derivatization chemistry. The new chemistry which will be reported on in greater detail elsewhere was carried out as part of a broad, general approach to the design of iodine accepting reagents capable of reacting in one step, under mild conditions with underivatized steroids and other clinically important analytes. Digoxin has six hydroxyl groups, many of which are available for reaction with electrophilic reagents. This fact, together with the known ability of a *para*-hydroxyphenyl group, such as that present in tyrosine, to act as an efficient iodine accepting moiety led to the synthesis of methyl-2-isocyanato-3(4′-trimethylsiloxyphenyl) propionate (Figure 11) trivially named TMS-tyrosine methyl ester isocyanate using organosilicon intermediates which uniquely served to convert the amine of tyrosine to an isocyanate while simultaneously blocking the *p*-hydroxyphenyl group. In the absence of the trimethylsilyl blocking group, the isocyanate would be susceptible to several nonproductive side-reactions.

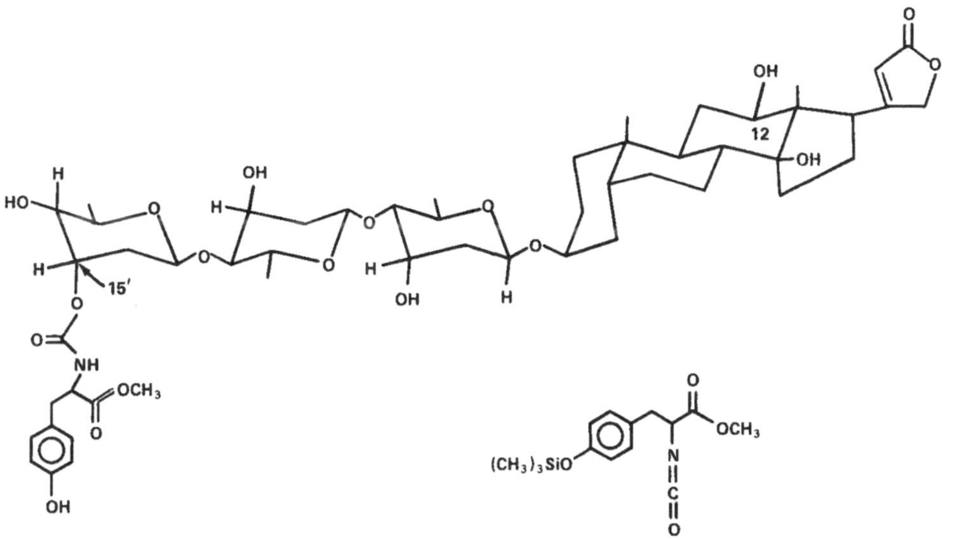

– Structural formulas of 15′–(TIME) carbamyldigoxin and novel isocyanate precursor.

FIGURE 11. Radiolabeling procedure for Centria® System assay ligands.

When the above isocyanate is allowed to react with digoxin at ambient temperature, a derivative, 15′-(TME)carbamyldigoxin (Figure 11), may be isolated as a pure compound. Radioiodination was carried out with Na ^{125}I using a modification of the Hunter and Greenwood chloramine T method. A high specific activity monoiodinated derivative is formed in excellent yield and is employed in the RIA in the presence of a small amount of surface active agent such as Tween® 80.

3. *Thyroxine Radioimmunoassay*

A standard curve was established that accommodates thyroxine concentrations between 1 and 30 μg/dℓ (Figure 12). Excellent linearization of the curve is achieved by the LOGIT-LOG transformation, which made this algorithm perfectly acceptable for data reduction.

Samples previously analyzed at the Steroid Research Institute of Montefiore Hospital, Bronx, N.Y. by radioimmunoassay were run. Assuming a normal range for thyroxine of 51 to 117 ng/mℓ, all serum samples showed the correct assigned pathologic or normal status when analyzed by the two differing procedures.

Table 3 gives an indication of the reproducibility obtainable with the prototype system. The first four lines show the reproducibility of the recalculated standard values of five consecutive runs. Thyroxine concentrations of control sera derived from simultaneously run standards are shown on the next four lines.

Standard data from previous runs can be stored in the microprocessor for use in calculating concentrations in serum in subsequent runs. This feature permits the loading of discs after the initial run with as many as 15 samples in duplicate, thus considerably increasing the throughput of the system. When intrarun precision was determined, data from a previous run was stored in the computer and retrieved for the calculation of precision in a second run in which all sample positions were filled with a specimen of pooled serum. Intrarun precision (CV) on the system has typically ranged from 3.5 to 5%; run-to-run precision was 7% when tested in our laboratories.

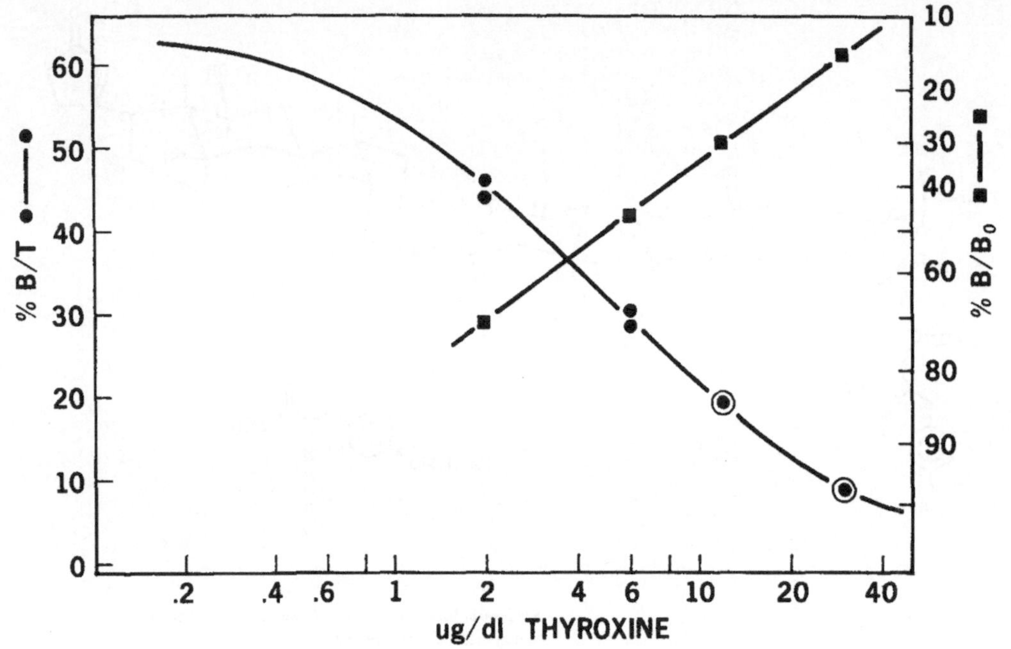

FIGURE 12. Standard curve for thyroxine radioimmunoassay. (From Ertingshausen, G., Shapiro, S. I., Green, G., and Zborowski, G., *Clin. Chem.*, 21, 1305, 1975. With permission.)

Table 3
RESULTS OF FIVE CONSECUTIVE RUNS OF A THYROXINE RADIOIMMUNOASSAY

Standard 1 (2μg/dℓ)	1.8±0.1	1.7±0.4	2.0±0.1	1.9±0.5	1.9±0.2
Standard 2 (6μg/dℓ)	6.4± 0.5	7.0±0.0	5.9±0.6	6.2±0.4	6.9±0.3
Standard 3 (12μg/dℓ)	12.4±0.5	12.3±2.6	11.4±0.0	13.0±0.0	13.4±0.5
Standard 4 (30μg/dℓ)	27.7± 1.9	27.8±2.0	30.2±0.0	28.3±1.0	28.0±3.1
Control 1	8.5	8.6	8.1	8.1	8.3
Control 1	9.1	8.0	8.1	8.1	7.8
Control 2	17.0	16.0	15.1	18.3	17.3
Control 2	15.9	17.1	9.9	17.1	15.9

From Ertinghausen, G., Shapiro, S. I., Green, G., and Zborowski, G., *Clin. Chem.*, 21, 1305, 1975.

a. Clinical Evaluation

The commercial version of the T_4 assay was later submitted to a clinical evaluation protocol in four hospital laboratories and the following results were obtained:

(1) Expected Values

The new Centria® T_4 RIA was clinically validated when compared to reference thyroxine radioassay methods used in four hospital laboratories. Patient samples were analyzed by the commercial T_4 RIA and selected in-house thyroid function tests.

A total of 457 patient samples were examined. Of these, 341 were determined to be euthyroid by reference test criteria. Statistical analysis of these results indicated an expected euthyroid range of 45 to 110 ng/mℓ for the Centria® T_4 RIA (Table 4).

Table 4
DETERMINATION OF NORMAL RANGE FOR CENTRIA® SYSTEM THYROXINE RADIOIMMUNOASSAY

	Number of samples analyzed		Centria® T_4 ng/mℓ		
Laboratory	Total	Normal	$\overline{X}$	S	$(\overline{X}-2S)$ to $(\overline{X}+2S)$
I	49	41	76.2	18.9	38—114
II	71	27	78.2	12.4	53— 103
III	118	100	72.3	18.0	36—108
IV	219	175	85.0	15.1	55—115
	Summary of euthyroid range				45—110

Table 5
CORRELATION DATA FOR CENTRIA® THYROXINE RADIOIMMUNOASSAY

		Linear regression analysis		
Evaluator	No. of clinical samples	Slope	Intercept	Correlation coefficient
A	195	0.878	5.754	0.944
B	118	0.991	8.898	0.933

Table 6
PARALLELISM DATA FOR THYROXINE RADIOIMMUNOASSAY

T_4 Concentration in ng/mℓ

	Sample 1		Sample 2		Sample 3	
Dilution factor #	Calc.	Observed	Calc.	Observed	Calc.	Observed
Original	—	41.73	—	92.97	—	173.1
1:2	20.35	21.52	46.89	44.05	86.55	84.51
1:3	13.91	14.00	30.99	28.71	57.7	52.08
1:4	10.47	10.41	23.24	21.02	43.02	36.18

(2) Specific Performance Characteristics

(a) Accuracy

1. Clinical Samples Comparison with Two Reference Methods: Clinical samples were analyzed by the Centria® System in parallel with the in-house reference T_4 RIA methods, routinely performed in two hospital RIA laboratories. The following results were established (Table 5).
2. Parallellism Study: In this study samples were diluted with T_4 RIA eluant and then analyzed for thyroxine. Results are summarized in Table 6.
3. Recovery Study: Recovery results are summarized in Table 7.

Table 7
RECOVERY DATA FOR THYROXINE RADIOIMMUNOASSAY

Conc. added T_4 ng/mℓ	Sample 1 Conc. calc.	Sample 1 Conc. observed	Sample 1 Percent recovery	Sample 2 Conc. calc.	Sample 2 Conc. observed	Sample 2 Percent recovery	Sample 3 Conc. calc.	Sample 3 Conc. observed	Sample 3 Percent recovery
0	—	43.41	—	—	95.17	—	—	163.6	—
30	73.41	72.57	97.2	125.17	130.6	118.1	193.6	196.0	108.0
60	103.41	105.4	103.3	155.17	168.1	121.6	223.6	228.7	108.5

Note: # Concentration in ng/mℓ.

Table 8a
PRECISION DATA FOR THYROXINE RADIOIMMUNOASSAY

Evaluator A — Within-Run variation

Control sera	$\bar{X}$	S	N	% CV
1	45.7	3.8	22	8.2
2	97.2	6.8	22	7.0
3	175.0	14.4	22	3.2

Table 8b
PRECISION DATA FOR THYROXINE RADIOIMMUNOASSAY

Evaluator A — Between-Run variation

Control sera	$\bar{X}$	S	N	% CV
1	37.0	2.8	22	7.4
2	65.2	3.2	22	4.9
3	171.7	11.8	22	6.9

Table 8c
PRECISION DATA FOR THYROXINE RADIOIMMUNOASSAY

Evaluator — Between-Run variation

Control sera	$\bar{X}$	S	N	% CV
1	41.7	3.5	26	8.4
2	91.7	4.8	26	5.3
3	164.0	11.1	26	5.7

Table 8d
PRECISION DATA FOR THYROXINE RADIOIMMUNOASSAY

Evaluator — Between-Run variation

Control sera	$\bar{X}$	S	N	% CV
1	37.3	2.9	15	7.8
2	66.3	8.5	15	12.9
3	163.7	11.3	15	6.9

Note: X = Mean T_4 concentration in ng/mℓ. S = Standard deviation. N = Number of replicate observations. CV = Coefficient of variation.

(b) Precision

The within-run and between-run variations of three control sera in each of the two clinical evaluator's laboratories are tabulated in Tables 8a to 8d.

(c) Specificity

The cross-reactivity with structurally similar molecules was studied and results are summarized in Table 9.

(d) Sensitivity

The sensitivity of this assay is determined to be 3.0 ng/mℓ. Sensitivity is defined as the minimum concentration of thyroxine which can be statistically distinguished from the zero calibration standard in the assay.

Table 9
SPECIFICITY DATA OF THYROXINE RADIOIMMUNOASSAY

Substance	Percent cross-reactivity
D-Thyroxine	100.0
DL-Thyroxine	100.0
3,3′,5,5′-Tetraiodothyroacetic Acid	24.2
3,3′,5-Triiodothyropropionic Acid	6.9
Triiodothyroacetic Acid	1.9
3,3′,5-Triiodo-DL-thyronine (free acid)	1.8
3,3′,5-Triiodo-D-thyronine	0.4
3-Iodo-L-tyrosine	$<10^{-2}$
3,5-Diiodo-L-thyronine	$<10^{-2}$

Note: % Cross-reactivity = c analyte ÷ c cross-reaction × 100 as measured from LOGIT-LOG plot of both analyte and its respective cross-reactant at 50% B_0 response level.

From Ertinghausen, G., Shapiro, S. I., Green, G., and Zborowski, G., *Clin. Chem.*, 21, 1305, 1975. With permission.

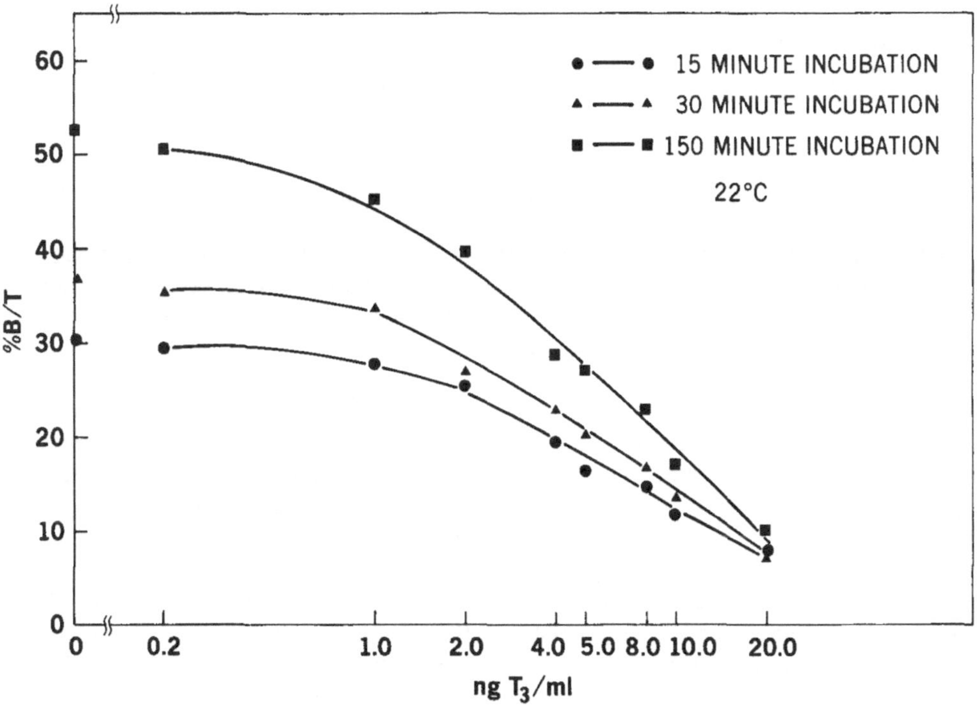

FIGURE 13. Standard curve for triiodothyronine radioimmunoassay obtained after three incubation times. (From Ertingshausen, G., Shapiro, S. I., Green, G., and Zborowski, G., *Clin. Chem.*, 21, 1305, 1975. With permission.)

4. Triiodothyronine Radioimmunoassay

To determine triiodothyronine by radioimmunoassay, we used the system's unique capability to perform kinetic, nonequilibrium assays. Figure 13 shows the standard curves obtained after 15 , 30 , and 150 min incubations. Because abnormally high T_3 concentrations appear to be of foremost clinical interest, sensitivity of the assay was reduced in favor of speed of analysis. If higher sensitivity is desired, incubation times

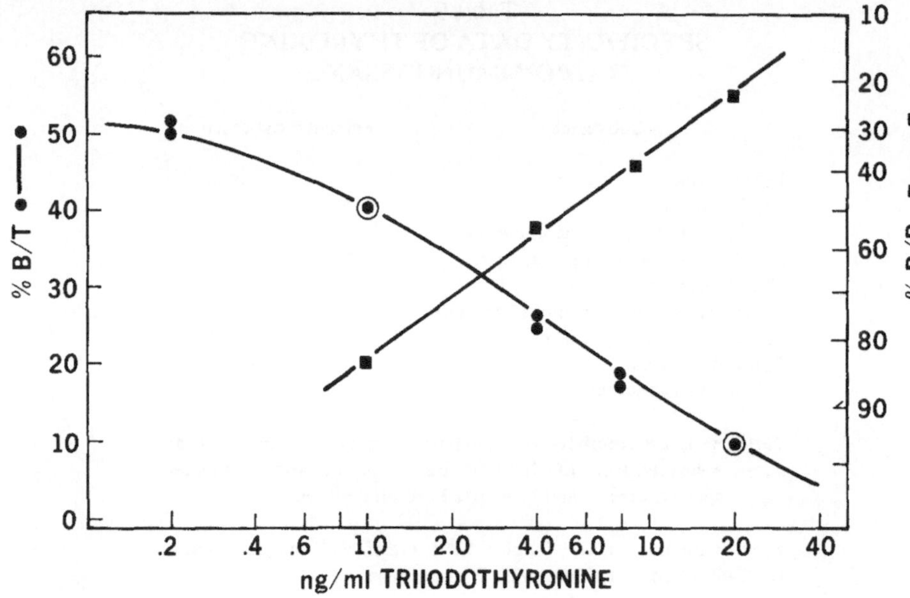

FIGURE 14. Standard curve for triiodothyronine radioimmunoassay with 30 min incubation. (From Ertingshausen, G., Shapiro, S. I., Green, G., and Zborowski, G., *Clin. Chem.*, 21, 1305, 1975. With permission.)

can easily be increased by removing the transfer disc from the incubator/separator after mixing and subjecting the incubated mixtures to separation after 2 hr have elapsed.

A standard curve covering concentrations between 0.2 and 20 ng of T_3/mℓ was established after 30 min incubation (Figure 14) and concentrations in serum were determined in a series of mostly pathologic samples. These have been analyzed by an equilibrium procedure performed at 4°C, with use of a second antibody to separate free from bound radiolabel.

Thirty-four samples were compared by establishing the regression equation $y = 0.95 \times - 0.12$, with a correlation coefficient of 0.97. These results indicated a bias in favor of higher values for our procedure but show an excellent correlation in view of the totally different assay approaches. Precision (CV) of the procedure between runs, determined on control samples at two different concentrations, was 9%; within-run precision was 7%.

The kinetic procedure uses the *tris*(hydroxymethyl)amino-methane buffer/Merthiolate reagent system originally suggested by Hüfner and Hesch.[8] We confirmed these authors' observation that antibody binding is essentially independent of Merthiolate concentration.

a. *Clinical Evaluation*

The commercial procedure which uses ANS to dissociate T_3 from TBG and requires a 2 hr incubation was later clinically validated. The following results were obtained:

(1) Expected Values

A group of 71 serum samples was analyzed by conventional T_4 RIA, and T_3 uptake methods. Those samples considered euthyroid by these methods exhibited a T_3 concentration range of 0.6 to 1.8 ng/mℓ when analyzed by the Centria® T_3 RIA method.

Table 10
CORRELATION DATA FOR CENTRIA® TRIIODOTHRYONINE RADIOIMMUNOASSAY

	Reference lab	Method 1	Method 2
N	44	82	31
r	0.93	0.95	0.96
slope	0.92	0.75	0.66
y-intercept	0.38	0.29	0.27

Note: N = number of samples
r = correlation coefficient
slope, y-intercept } calculated from linear regression

Table 11
PARALLELISM DATA FOR TRIIODOTHYRONINE RADIOIMMUNOASSAY

T_3 Concentrations in ng/m*l*

Dilution factor	Sample 1 Calc.	Sample 1 Obs.	Sample 2 Calc.	Sample 2 Obs.	Sample 3 Calc.	Sample 3 Obs.
Original	—	2.25	—	3.10	—	5.26
1 : 2	1.13	1.38	1.55	1.84	2.63	3.25
1 : 4	0.56	0.65	0.78	0.91	1.31	1.68

Table 12
RECOVERY DATA FOR TRIIODOTHYRONINE RADIOIMMUNOASSAY

Conc. added in ng/m*l*	Sample 1 Conc. calc.*	Sample 1 Conc. obs.	Sample 1 Percent recovery	Sample 2 Conc. calc.	Sample 2 Conc. obs.	Sample 2 Percent recovery
		0.49			0.22	
2.00	2.49	2.53	102	2.22	2.01	90
4.20	4.69	4.07	85	4.42	4.45	101

* Concentration in ng/m*l*.

(2) Specific Performance Characteristics

(a) Accuracy

1. Clinical Sample Comparisons with Reference Methods: Centria® T_3 RIA concentrations in clinical samples were compared to those obtained by an independent reference laboratory and with two other commercial methods. Results of the linear regression analyses are presented below for Centria® vs. reference (Table 10).
2. Parallelism Study: Parallelism reults are summarized in Table 11.
3. Recovery Study: Table 12 shows the results of recovery studies.

Table 13
PRECISION DATA FOR TRIIODOTHYRONINE RADIOIMMUNOASSAY

Control sera	Within-Run variation $\overline{X}$	S	N	CV	Between-Run variation $\overline{X}$	S	N	CV
1	1.79	0.083	19	4.60	1.81	0.062	9	3.40
2	3.39	0.129	20	3.80	3.27	0.151	9	4.60
3	1.09	0.050	20	4.60	1.21	0.082	9	6.80
4	2.78	0.150	20	5.40	2.93	0.100	9	3.40
5	0.75	0.058	20	7.60	0.89	0.118	9	13.2 0
6	1.43	0.108	20	7.50	1.59	0.136	9	8.60
7	2.92	0.142	20	4.90	3.23	0.093	9	2.90
8	1.20	0.088	18	7.30	1.19	0.075	9	6.30
9	2.44	0.096	20	3.90	2.59	0.106	9	4.10

Note: $\overline{X}$ = mean T_3 concentration in ng/mℓ;
S = standard deviation;
N = number of replicate observations;
CV = coefficient of variation in percent.

Table 14
SPECIFICITY DATA FOR TRIIODOTHYRONINE RADIOIMMUNOASSAY

Substances	Percent cross-reactivity
D-Triiodothyronine	12.000
L-Thyroxine	0.236
D-Thyroxine	0.300
Tetraiodothyroacetic Acid	6.900
Monoiodotyrosine	$< 10^{-3}$
Diiodotyrosine	$< 10^{-4}$
Sodium Salicylate	$< 10^{-4}$
5,5-Diphenyl-Hydantoin	$< 10^{-4}$
Phenylbutazone	$< 10^{-4}$
Methimazole	$< 10^{-4}$
6-*N*-Propyl-2-Thio-Uracil	$< 10^{-4}$

Note: Percent Cross-Reactivity = cT_3/c cross-reactant × 100 observation made at 50% B_o response level in the calibration curves of both T_3 and respective cross-reactants.

(b) Precision

The within-run and between-run variations found in analyses of nine control sera are tabulated in Table 13.

(c) Specificity

Cross-reactivity data are summarized in Table 14.

(d) Sensitivity

The sensitivity of this assay is 0.2 ng/mℓ. Sensitivity is defined as the minimum concentration of T_3 which can be statistically distinguished from the zero calibration standard in the assay.

5. T_3 Uptake Test

Use of Sephadex® G-25 in the T_3 uptake test was first proposed by Lissitzky et al.[9] Of all the separation media suggested for this test, Sephadex® G-25 subjects the incubated mixture to the mildest separation process, which is essentially based on dialysis and does not affect the equilibrium.

Results obtained with our procedure on 40 clinical samples were compared to results given by the manual Trilute® test (Ames. Co., Elkhart, Ind. 46514). A linear regression equation of $y = 0.87x + 9.33$ and a correlation coefficient of $r = 0.87$ were observed. Most of the results are clustered in and around the normal range, which usually leads to poor correlation data. The means of all data were, however, very close: 42.1% for the Trilute® test and 43.9% for the Centria® System.

a. *Clinical Evaluation — Expected Values*

(1) Percent T_3U Retention

The Centria® System T_3U radioassay was clinically validated in four laboratories, including three large hospitals and an in-house facility. Patient samples were analyzed by the Centria® System and selected reference thyroid function tests.

A total of 235 patient samples were examined. Of these, 133 were determined to be euthyroid by reference test criteria. Statistical analysis of these results indicated an expected range of 38 to 56% T_3U retention for the Centria® System T_3 uptake assay. Values less than 38% appear in the hypothyroid range, and values greater than 56% appear in the hyperthyroid range.

(2) Free Thyroxine Index

Free thyroxine concentration may be quantitated directly utilizing an equilibrium dialysis procedure. This method is not practical for routine testing in most hospital laboratories.

A free thyroxine index (FTI) is one alternative which provides the clinician with a rapid in vitro assessment of thyrometabolic status. The FTI is calculated by multiplying the total thyroxine concentration (expressed in μg/dl) by the percent T_3U retention (expressed as a decimal). From analysis of 133 euthyroid patient results, a FTI expected range of 2.1 to 5.2 is obtained.

A summary of these results is shown in Figure 15.

b. *Specific Performance Characteristics*

(1) Precision

The within-run variation for the three control sera is tabulated in Table 15a. The between-run variation for the three control sera is tabulated in Table 15b.

6. Digoxin Radioimmunoassay

The digoxin procedure was originally developed with a radio-iodinated digoxin/tyrosine methylester as supplied by Burroughs-Wellcome Co. The standard curve covered the range of concentrations between 0.2 and 8 ng of digoxin per milliliter and was adequately linearized by the LOGIT-LOG transformation (Figure 16). Intrarun precision was determined to be 3.5 to 4%, while interrun precision for the prototype was 9%.

Fifty-eight clinical samples, previously analyzed at Methodist Hospital, Indianapolis, by a procedure in which a tritiated digoxin ligand and dextran-coated charcoal are used in the separation step, were reassayed on the Centria® system. Regression analysis ($y = 1.03x - 0.17$; $r = 0.94$) revealed good agreement. In the toxic range — i.e., above concentrations of 2 ng/mℓ digoxin — the two methods disagreed for only one sample.

LABORATORY	NUMBER OF SAMPLES ANALYZED		% T_3U RETENTION			FREE THYROXINE INDEX	
	TOTAL	NORMAL	$\bar{X}$	S	HISTOGRAM	$\bar{X}$	S
I	71	27	46.0	4.1	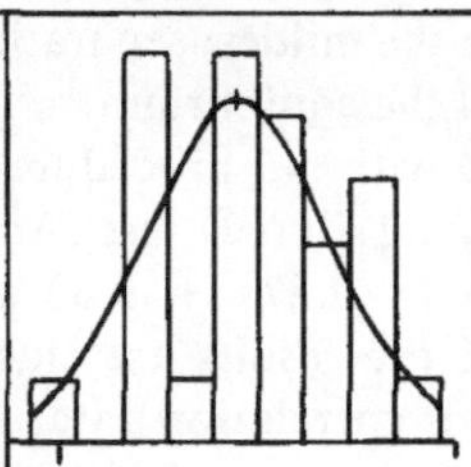	3.6	0.7
II	48	40	47	4.2	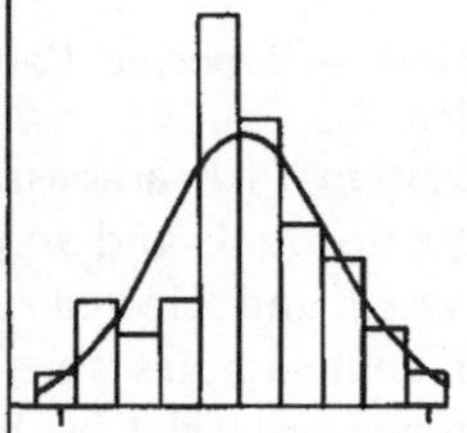	3.6	1.0
III	63	20	48	5.4	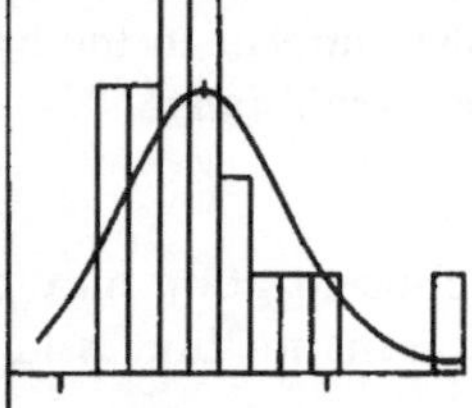	3.4	0.5
IV	53	46	47	4.7	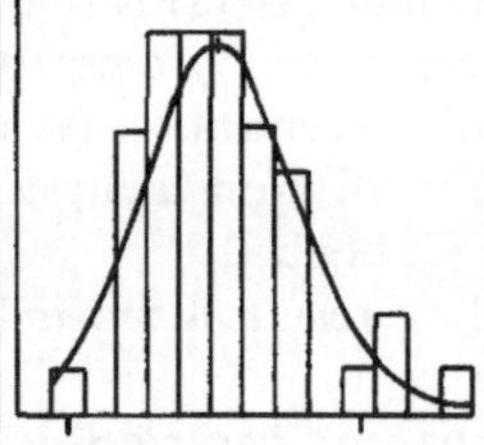	3.8	0.7
SUMMARY OF EXPECTED RANGES			38-56			2.1-5.2	

FIGURE 15. Determination of expected range for Centria® System % T_3U Retention Assay in four laboratories.

a. Clinical Evaluation

In the commercial procedure the iodinated digoxin/tyrosine methylester was replaced with a proprietary radiolabeled digoxin ligand.[7]

(1) Expected Values

Clinical studies have defined therapeutic (1.4 ± 0.7 [SD] ng/m*l*) and toxic (3.7 ± 1.0 [SD] ng/m*l*) levels of digoxin concentrations. Clinical samples from digitalized patients in two large hospitals were analyzed by the System's digoxin RIA method. The frequency distributions of these two hospital populations are shown in Figure 17.

Table 15a
PRECISION DATA FOR % T_3 UPTAKE TEST

Within-run variation

Control	$\overline{X}$	S	N	CV
I	40.9	1.2	32	2.9
II	50.4	1.8	32	3.5
III	69.5	2.5	32	3.7

Table 15b
PRECISION DATA FOR % T_3 UPTAKE TEST

Between-run variation

Control	$\overline{X}$	S	N	CV
IV	39.7	1.2	20	3.1
V	43.1	2.2	20	5.1
VI	51.5	2.7	20	5.2

Note: $\overline{X}$ = percent T_3U retention;
S = standard deviation;
N = number of replicate observations;
CV = coefficient of variation.

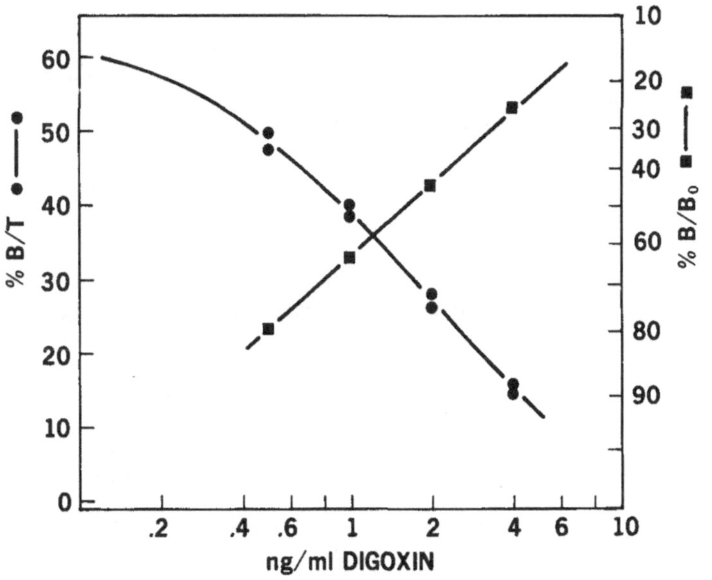

FIGURE 16. Standard curve of digoxin radioimmunoassay. (From Ertingshausen, G., Shapiro, S. I., Green, G., and Zborowski, G., *Clin. Chem.*, 21, 1305, 1975. With permission.)

(2) Specific Performance Characteristics
(a) Accuracy

1. Clinical Sample Comparison With Two Reference Methods: Clinical samples

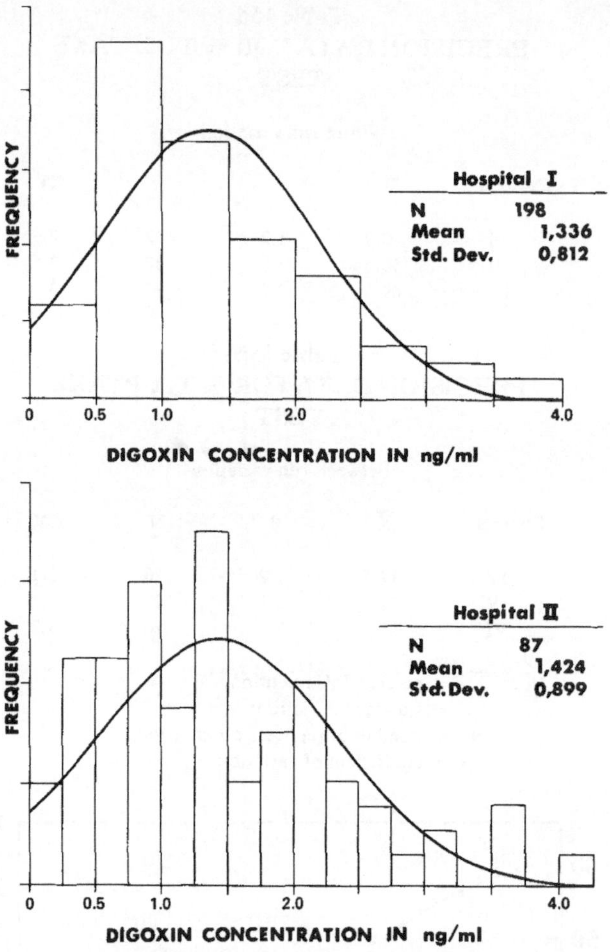

FIGURE 17. Digoxin concentration in digitalized patients in two hospitals.

Table 16
CORRELATION DATA FOR CENTRIA® SYSTEM DIGOXIN ASSAY

		Linear regression analysis		
Evaluator	No. of clinical samples analyzed	Slope	Intercept	Correlation coefficient
A	188	1.143	0.176	0.916
B	87	1.054	0.136	0.958

were analyzed by the Centria® System method in parallel with the in-house reference digoxin RIA methods, routinely performed in two hospital RIA laboratories. The following results were obtained (Table 16).

2. Parallelism study (Table 17).
3. Recovery study (Table 18).

Table 17
PARALLELISM DATA FOR DIGOXIN RADIOIMMUNOASSAY

Dilution factor	Digoxin concentration in ng/ml					
	Sample 1		Sample 2		Sample 3	
	Calc.	Obs.	Calc.	Obs.	Calc.	Obs.
Original	—	0.80	—	1.84	—	3.34
1 : 2	0.40	0.40	0.92	0.92	1.67	1.70

Table 18
RECOVERY DATA FOR DIGOXIN RADIOIMMUNOASSAY

	Sample 1	Sample 2	Sample 3
Original conc.[a]	0.44	0.44	0.44
Conc. added	0.49	0.98	3.92
Conc. observed	1.00	1.46	4.45
Percent recovery	114%	104%	102%

[a] Concentration in ng/ml.

Table 19
PRECISION DATA FOR DIGOXIN RADIOIMMUNOASSAY

Control sera	Within-run variation			
	$\bar{X}$	S	N	CV
1	1.06	0.06	21	5.7%
2	1.72	0.10	22	5.8%

Note: $\bar{X}$ = mean digoxin concentration in ng/ml;
S = standard deviation;
N = number of replicate observations;
CV = coefficient of variation.

(b) Precision

The within-run variation of each concentration found for the two control sera are tabulated in Table 19. The between-run variation of each concentration found for two hospital RIA laboratories over the clinical evaluation period are tabulated in Table 20.

(c) Specificity

See Table 21.

(d) Sensitivity

The sensitivity of this assay is 0.1 ng/ml. Sensitivity is defined as the minimum concentration of digoxin which can be statistically distinguished from the zero calibration standard in the assay.

7. Insulin Radioimmunoassay

Insulin — Figure 18 shows an insulin standard curve which was obtained after a 20

Table 20
BETWEEN RUN VARIATION FOR DIGOXIN RADIOIMMUNOASSAY

Control sera	Hospital I				Hospital II			
	$\bar{X}$	S	N	CV	$\bar{X}$	S	N	CV
1	0.63	0.08	7	13.0%	0.61	0.07	17	12.0%
2	1.64	0.11	10	6.8%	1.55	0.10	17	6.6%
3	2.87	0.10	10	3.5%	2.84	0.21	17	7.6%
4	1.10	0.09	14	8.2%	1.06	0.10	17	9.8%
5	2.99	0.22	14	7.4%	2.96	0.39	17	13.0%

Table 21
SPECIFICITY DATA FOR DIGOXIN RADIOIMMUNOASSAY

Substances	Percent cross-reactivity
Deslanoside	55.0
Digitoxin	6.8
Digitoxigenin	5.5
Spironolactone	0.1
Ouabain	0.1
Testosterone	< 0.001
Progesterone	< 0.001
Cholesterol	< 0.001
Prednisone	< 0.001

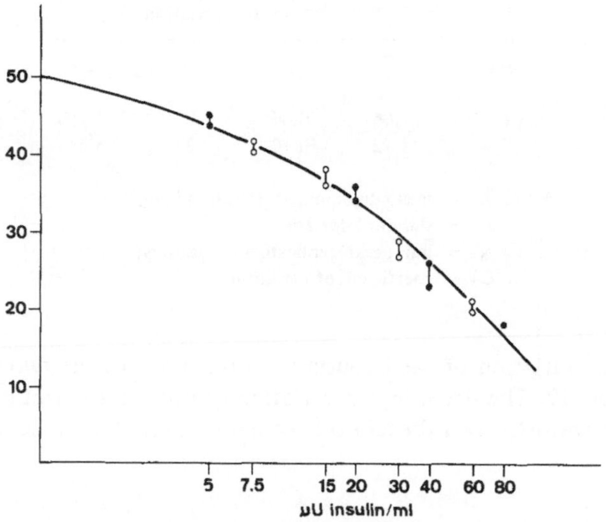

FIGURE 18. Standard curve of insulin radioimmunoassay. (From Green, G., Dyce, D., Gimovsky, A., and Lo, D. H., *Clin. Chem.*, 22, 1510, 1976. With permission.)

hr incubation at 4°C.[6] The indicated sensitivity is < 5 mU of insulin per liter. In addition, Figure 17 shows that the criterion of parallelism is satisfied in this assay, because dilutions of a patient's serum sample exhibit points that fall on the standard curve. Insulin values obtained by an outside laboratory using the Amershan/Searle Insulin

Table 22
CORRELATION DATA FOR INSULIN RADIOIMMUNOASSAY

Laboratory	No. of samples analyzed	Linear regression analysis: Slope	Intercept	Correlation coefficient
I	67	0.628	13.076	0.914
II	54	0.675	6.201	0.973

Table 23
PARALLELISM DATA FOR INSULIN RADIOIMMUNOASSAY

Dilution factor	Insulin concentration in μU/mℓ: Sample 1 Calc.	Sample 1 Observed	Sample 2 Calc.	Sample 2 Observed	Sample 3 Calc.	Sample 3 Observed
Original		21.2		53.4		117.3
1:2	10.6	9.9	26.7	23.4	58.7	56.3

Tables 22 and 23 from Ertinghausen, G., Shapiro, S. I., Green, G., and Zborowski, G., *Clin. Chem.*, 21, 1305, 1975. With permission.

Radioimmunoassay Kit were also determined on a Centria® analyzer. A least-squares linear regression equation of y + 1.037x + 2.9 was obtained for 98 patients' samples (y = Centria® x = reference laboratory), with a correlation coefficient of 0.965. The means determined for all 98 samples were 43.0 mU of insulin per liter (reference laboratory) and 47.5 mU of insulin per liter (Centria®).

a. *Clinical Evaluation*

The commercial assay for insulin uses Sephadex® DEAE-25 columns.

(1) Expected Values

(a) Fasting

Normal fasting (overnight) levels are 21 ± 15 μU/mℓ. This range may be elevated in obese individuals and reduced in patients on carbohydrate restricted diets. Eighteen normal fasting clinical samples were analyzed by the System's Insulin RIA procedure. An immunoreactive insulin concentration range of 5.2 to 24.3 μU/mℓ was established.

(2) Specific Performance Characteristics

(a) Accuracy

1. Clinical Sample Comparison: Centria® immunoreactive insulin concentrations were determined in 115 clinical samples in two hospital RIA laboratories. The following results were established (Table 22).
2. Parallelism Study. (Table 23.)
3. Recovery Study. (Table 24.)

(b) Precision

The within-run variations in concentrations determined in five control sera are tabulated in the following Table. (Table 25).

Table 24
RECOVERY DATA FOR INSULIN RADIOIMMUNOASSAY

	Sample 1	Sample 2	Sample 3
Original conc.	11.2	32.5	56.4
Conc. added	143.3	130.9	143.3
Conc. observed	156.9	161.0	202.8
Percent recovery	101.6%	98.2%	102.2%

Table 25
PRECISION DATA FOR INSULIN RADIOIMMUNOASSAY

	Within-run variation			
Control sera	$\overline{X}$	S	N	CV
1	13.9	2.8	20	20.2%
2	34.3	3.0	20	8.7%
3	103.1	5.5	20	5.4%

C. Procedures for Large Polypeptide Molecules

The Centria® System using a gel filtration column separation was less easily adapted to the separation of macromolecules. Gel filtration procedures with Sephadex® G-50 or G-75 had to be ruled out because the mechanical strength of the swelled material was insufficient for use in the centrifugal field. A separation procedure using Sephadex® DEAE-50, however, proved successful in a number of selected cases.

The procedure was successfully applied to radioimmunoassays for insulin, somatotropin[6] (human growth hormone), and choriomammotropin[10] (human placental lactogen).

1. Ion-Exchange Column Preparation and Use

Swelling — Phosphate buffer (pH 6.6, 0.2 mol/ℓ; was) used for swelling. Ten to fifteen grams of diethylaminoethyl-Sephadex® A-50 was suspended in 300 mℓ of the phosphate buffer, covered, and swelled overnight or until the slurry was thin enough to pipette. The slurry was degassed thoroughly before use.

Columns — Full-sized Sarstedt® columns (Walter Sarstedt, Inc., Princeton, N.J. 08540) were used. The bottom plug was a Porex® chromic acid-treated hydrophilic plug, prewet with the pH 6.6 phosphate buffer, and drained.

Elution — Columns were eluted in the centrifugal field with phosphate buffer (pH 7.0, 25 mmol/ℓ). The columns were never commercialized. Instead a more versatile separation procedure involving the use of immobilized second antibody was developed and successfully adapted to the Centria® System.

2. Universal Solid Phase Double Antibody Procedure

We were unable to provide a general column separation procedure for macromolecules until Henry and Meriadec investigated the potential of a solid phase double antibody technique for use with our separation columns. We had previously established that a hydrophobic disc placed at the bottom of a separation column was able to retain the liquid in a column containing swollen gel with entrapped antibody while spinning at 100 r/min in the Centria® separator/incubator. The hydrophobic barrier could be broken once a certain amount of liquid had been added to the gel with the eluant pump.

Instead of applying the gel-bound antibody as primary antibody in the assay the

Table 26
PRECISION OF TSH STANDARD CURVE

	r	Slope	Intercept	(B/B_o) = 50%	(B/B_o) % Standard in μU/mℓ: 2	4	10	20	40	100
$\overline{X}$	0.997	1.125	3.35	18.96	92.9	86.1	68.2	47.5	28.8	15.4
σ	0.001	0.05	0.18	1.65	1.45	1.23	2.04	2.95	2.90	1.94
CV%	0.1	4.4	5.3	9	1.6	1.4	3	6.2	10.1	12.6

new separation techinque uses Sepharose-immobilized anti-IgG as a means to separate free from bound radioligand. The dry Sepharose swells once in contact with the RIA incubate and reacts with all the antibody including the antibody-bound ligand. The Sepharose bound anti-IgG is added in tablet form to the column fitted with a hydrophobic frit. The overall procedure is thus convenient and simple to handle. In order to count the bound label the System had to be modified such that the columns could be lifted from the incubator/separator without the test tubes and counted on the counter/computer.

3. Radioimmunoassay for Thyroid Stimulating Hormone (TSH)-
developed by BioSystèmes, S.A., an affiliate of Union Carbide France, S.A.

a. *Performance Characteristics*

1. Pre-incubation time: 25 hr in the disc
2. Incubation time: 2 hr in the disc, 10 min in the column.
3. Temperature: 37°C
4. Sample volume: 50 μℓ
5. Separation procedure: second antibody on solid phase (tablets)

(1) Sensitivity

(B/B_o) = 95% corresponds to 1.5 μU/mℓ.

(2) Precision

1. Standard curve. N = 15 (Table 26)
2. Precision intra-run

n° pool	PG
mean (μU/mℓ)	6.43
coefficient of variation	1.7%
number of determinations	9

3. Precision inter-run

n° pool	PG
mean (μU/mℓ)	6.78
coefficient of variation	7.6%
number of determinations	23

(3) Accuracy

Dilution curves see Table 27.
Recovery tests see Table 28.

(4) Specificity

Saturated with HCG.
No cross-reaction at (B/B_o) = 50% with HCG, LH, FSH.

Table 27
DILUTION CURVE — TSH RADIOIMMUNOASSAY

	Serum A μU/mℓ		Serum B μU/mℓ	
Dilution	Theor	Exp	Theor	Exp
1/1		61.24		78.80
1/2	30.62	32.69	39.40	41.19
1/4	15.31	15.33	19.70	20.61
1/8	7.60	7.15	9.85	10.47
1/16	3.83	3.84	4.93	5.55
1/32			2.46	3.02
coeff. corr.	0.999		1	

Table 28
RECOVERY RATE — TSH RADIOIMMUNOASSAY

	Serum A			Serum B			Serum C		
TSH added μU/mℓ	Theor μU/mℓ	Exp μU/mℓ	Rec %	Theor μU/mℓ	Exp μU/mℓ	Rec %	Theor μU/mℓ	Exp μu/mℓ	Rec %
Serum		6.66			3.32			2.24	
0	3.33	4.75	143	1.66	1.60	96.4	1.12	0.71	
1	4.33	5.22	121	2.66	2.39	89.8	2.12	1.67	80
2	5.33	5.52	104	3.66	3.46	94.5	3.12	2.66	85
5	8.33	8.93	107	6.66	5.80	87.0	6.12	5.54	91
10	13.33	13.98	105	11.66	11.02	94.5	11.12	10.67	96
20	23.33	26.92	115	21.66	23.39	107.9	24.12	20.73	86
50	53.33	60.72	114	51.66	51.00	98.7	51.12	48.85	96
Coeff. Corr.		0.999			0.998			0.999	

(5) Precision

Slope $\overline{X}$ = 1.125, CV = 4.4%

(B/B_o) = 50% corresponds to: X = 19 μU/mℓ, CV = 1.7%

(6) Adjustment of Standards

The standards are calibrated with standard MRC 68/38.

(7) Expected Values

See Figure 19.

4. Radioimmunoassay for α-Fetoprotein (AFP) Developed by BioSystèmes, S.A., an Affiliate of Union Carbide France, S.A.

a. Performance Characteristics

Incubation time: 2 hr in damp atmosphere

Temperature: room temperature

Sample volume: 50 μl

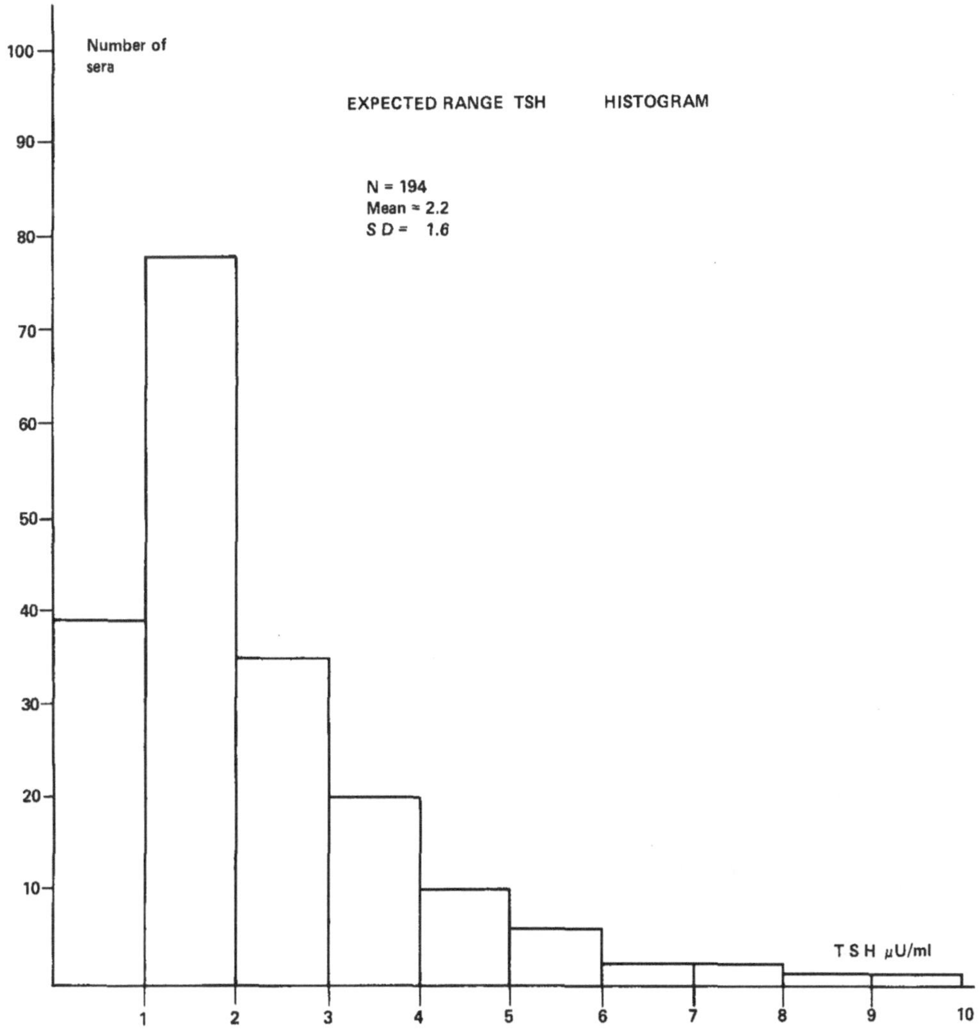

FIGURE 19. Determination of expected range for TSH assay in an evaluator's laboratory.

Table 29
INTERUM PRECISION FOR AFP ASSAY

r	Slope	Intercept	$(B/B_o) = 50\%$	Standards in ng/mℓ 10	20	50	100	300
$\bar{X}$ 0.998	−1.03	4.64	90.3	90.7	82.5	54.3	46.7	23.0
CV 0.01%	5	6.2	9.2	2.3	2.8	3.9	5.2	6.6

(1) Separation Procedure: Double Antibody Tablet

1. Sensitivity — $(B/B_o) = 95\% \rightarrow 4.5$ ng/mℓ
2. Precision intra-run (Table 29) — Standard curve (N = 45)
3. Precision inter-run

n° pool	4	5	6
mean (ng/mℓ)	21.8	134.5	7.5
coeff. of variation (%)	8.5	4.9	12.3
No. of determinations	10	10	10

Table 30
PARALLELISM FOR AFP ASSAY

Serum dilution	Serum 1	Serum 2 (ng/mℓ)	Serum 3 (ng/mℓ)
1/1	—	280	293.5
1/2	214.4	144	152.5
1/4	121.6	71.7	72.1
1/8	64.5	33.8	38.0
1/16	35.8	17.0	17.3
1/32	18.3	—	11.2
1/128	5.4	—	—

Table 31
RECOVERY DATA FOR AFP ASSAY

AFP added	Serum		Serum 2		Serum 3	
ng/mℓ	ng/mℓ	% rec	ng/mℓ	% rec	ng/mℓ	% rec
0	6		4		7.5	
20	26.3	101	25	104	26.8	97
50	56.6	101	54	100	51.1	89
100	105.3	99	110	105	100.2	93
300	298.9	98	300	99	285.8	93

4. Precision inter-run

n° pool	4	5	6
mean (ng/mℓ)	25.8	152.7	9
coeff. of variation (%)	14	7.6	27
No. of determinations	73	73	73

(2) Accuracy (Parallelism)

1. Dilution curves of 3 elevated sera: (Table 30)
2. Recovery Tests: (Table 31)
3. Specificity: no cross-reaction with bovine and human serum albumin
4. Normal Range: determined with 217 sera, presumed to be normal, from blood donors

   $\bar{X} = 6.7$ ng/mℓ

   $\sigma = 3.5$
5. Expected values (Figure 20)

5. Radioimmunoassay for Ferritin-developed by BioSystèmes, S.A., an Affiliate of Union Carbide France, S.A.

a. Performance Characteristics

Incubation time: 1 hr in the disc, 10 min in the column
Temperature: room temperature
Sample volume: 50 μl
Separation procedure: second antibody on solid phase (tablets)
Contents are sufficient for two discs

(1) Sensitivity

$(B/B_o) = 95\%$, corresponds to 4,5 ng/mℓ

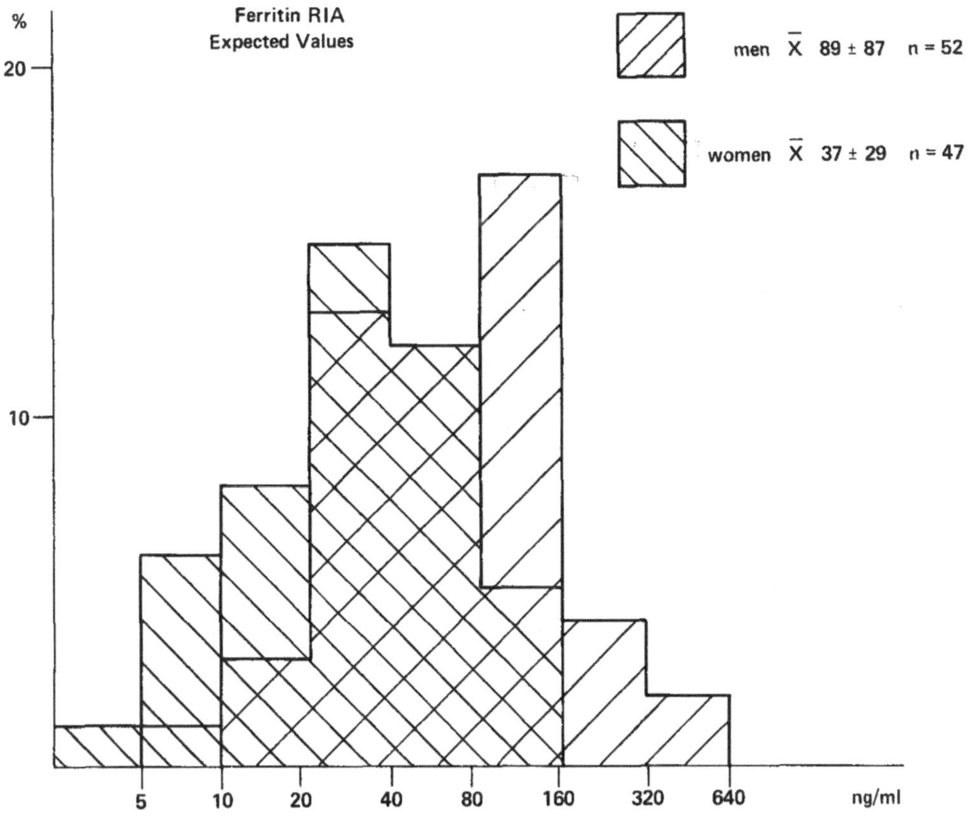

FIGURE 20. Determination of expected ranges for ferritin in evaluator's laboratory separately for women and men.

Table 32
PRECISION OF FERRITIN STANDARD CURVE

	r	Slope	Intercept	$(B/B_o) = 50\%$	(B/B_o) % Standard in ng/mℓ 10	20	30	40	50	60
$\bar{X}$	0.998	1.025	4.54	83.8	90.0	81.3	62.9	43.9	28.8	14.4
σ	0.0007	0.04	0.21	7.1	1.73	1.9	2.9	2.5	1.4	1.8
CV%	0.08	4.1	4.7	8.5	1.9	2.4	4.6	5.7	4.9	10.9

(2) Precision (Table 32)

1. Standard curve: N = 10
2. Precision intra-run:

n° pool	PG	PN
mean (ng/mℓ)	16.8	162
variation coefficient (%)	14.5	3.4
No. of determinations	10	10

3. Precision inter-run:

n° pool	PG
mean (ng/mℓ)	17.4
variation coefficient	14.5
No. of determinations	10

Table 33
DILUTION CURVES FERRITIN RIA

	Serum 1 ng/mℓ		Serum 2 ng/mℓ		Serum 3 ng/mℓ	
Dilution	Exp	Theor	Exp	Theor	Exp	Theor
1/1	272		208		158	
1/2	147	136	110	104	87	79
1/4	67	68	54	52	44	40
1/8	32	34	25	26	19	20
1/16	17	17	14	13	8.2	9.9
1/32	8.2	8.5	6.5	6.5	5.5	5.0

Table 34
RECOVERY DATA FERRITIN RIA

FH added	Serum 1		Serum 2		Serum 3	
ng/mℓ	ng/mℓ	% rec	ng/mℓ	% rec	ng/mℓ	% rec
0	27	111	2717	97	25	98
5	26	87	31.5	94	24	81
10	31	89	34	88	31	87
25	45	91	52	97	47	94
50	77	103	69	87	74	99
100	130	104	129	100	119	95
250	270	98	267	96	246	89

(3) Accuracy

Dilution curves see Table 33.
Recovery test see Table 34.

(4) Precision

Slope $\overline{X}$ = 1.025, CV = 4.1%
(B/B_o) = 50% corresponds to $\overline{X}$ = 83.8 ng/mℓ, CV = 8.5%

REFERENCES

1. **Anderson, N. G.**, Computer interfaced fast analyzers, *Science*, 166, 317, 1969.
2. **Ertingshausen, G., Shapiro, S. I., Green, G., and Zborowski, G.**, Adaptation of a T_3-uptake test and of radioimmunoassays for serum digoxin, thyroxine, and triiodothyronine to an automated radioimmunoassay system - "CENTRIA", *Clin. Chem.*, 21, 1305, 1975.
3. **Smith, T. W., Butler, V. P., and Haber, E.**, Characterization of antibodies of high affinity and specificity for the digitalis clycoside digoxin, *Biochemistry*, 9, 331, 1970.
4. **Vaitukaitis, J., Robbins, J. B., Nieschlag, E., and Ross, G. T.**, A method for producing specific antisera with small doses of immunogen, *J. Clin. Endocrinol. Metab.*, 33, 988, 1971.
5. **Gharib, H., Ryan, R. J., Mayberry, W. E. and Hockert, T.**, Radioimmunoassay for triiodothyronine (T_3). I. Affinity and specificity of the antibody for T_3, *J. Clin. Endocrinol. Metab.*, 33, 509, 1971.
6. **Green, G., Dyce, D., Gimovsky, A., and Lo, D. H.**, Automated approach to radioimmunoassays of somatrophin (human growth hormone) and insulin, *Clin. Chem.*, 22, 1510, 1976.
7. **Eisenhardt, W. A., Hedaya, E., and Theodoropulos, S.**, U.S. Patent 4,115,539, 1978.
8. **Hüfner, M. and Hesch, R. D.**, A comparison of different compounds for TBG-blocking used in radioimmunoassay, *Clin. Chem.*, 44, 101, 1973.
9. **Lissitzky, S., Bismuth, J., and Rolland, M.**, Séparation des composés iodés du serum et de la thyröide par filtration sur gel de dextran (Sephadex), *Clin. Chim.*, 7, 183, 1962.
10. **Castro, A., Prieto, J., Wunsch, C., Ertingshausen, G.**, Automated radioimmunoassay for choriomammotropin (human placental lactogen), *Clin. Chem.*, 22, 1655, 1976.

Chapter 2

THE MICROMEDIC® CONCEPT 4® AUTOMATIC RADIOASSAY

W. G. Lawn

TABLE OF CONTENTS

I. INTRODUCTION

In the past several years radioassay tests have played an increasingly important role in the diagnosis of disorders of the thyroid, endocrine system, and bloodforming organs, as well as in the detection of cancer, hepatitis, and drug addiction. However, a major obstacle to the greater use of radioassay tests has been the labor-intensive and time-consuming nature of manual assay procedures.

Micromedic® Systems, recognizing the need and benefits of automation in the radioassay laboratory, has developed several different instrument systems which tailor the degree of automation to the customer's needs. The CONCEPT 4® Automatic Radioassay, together with the complementary AUTOPAK® Test Delivery System, offers the ultimate in radioassay automation. Basically, CONCEPT 4® is a sample-to-answer radioassay system; i.e., it is a high-precision radioassay system that totally automates all functions from pipetting of samples and reagents to calculation and printout of final patient results. All of this is done without human intervention. Fundamentally, it is designed for those laboratories requiring high throughput and performance without the need for additional space and personnel.

In this chapter the operation, features, and benefits to the user of the CONCEPT 4® Automatic Radioassay will be discussed, together with the complementary AUTOPAK® Test Delivery System chemistries.

II. THE CONCEPT 4® AUTOMATIC RADIOASSAY

A. Overall Description

1. Instrument

The specifications for the CONCEPT 4® Automatic Radioassay are shown in Table 1. An overview of the instrument itself is shown in Figure 1.

The utility requirements for the instrument are quite simple:

1. No special electrical wiring is needed since CONCEPT 4® draws relatively little power from a standard laboratory 115 VAC electrical circuit.
2. No plumbing or water supply hookup is required.

CONCEPT 4® occupies relatively little laboratory space. The primary reasons for this are the miniaturization of the electronics and the use of 8 × 50 mm, rather than 12 × 75 mm, tubes in the chemistries used on this system (see Section II., A., 2.).

The instrument itself consists basically of five stations: sample pipetting, reagent

Table 1
SPECIFICATIONS

Sample Preparation			
Sample loading zone capacity			20 racks (200 tubes)
Assay loading zone capacity			20 racks (200 tubes)
Sample volume range			2—400 μl
Reagent volume range			100—1000 μl
Sample/Reagent pipetting rate			1 rack per min
Sample/Reagent mixing time			2 sec
Incubation			
Incubator capacity			200 tubes
Time setting			minimum of 1 min per assay rack to 999.0 min
Temperature setting			ambient, 37°, 45°C
Temperature regulation			0.5°C
Aspirate and Rinse			
Wash volume			1.4 ml
Counter			
Number of detectors			2
Background			< 75cpm
Counting time range			0.1 to 25 min
Sample changeover time			2 sec
Loading zone capacity			60 racks (600 tubes)
Post-counter capacity			60 racks (600 tubes)
Dimensions			
Width	62 in.	157.5 cm	
Depth	32 in.	81.3 cm	
Height	61 in.	155 cm	
Weight	765 lb	347 kg	
Power Requirements			
800 VA			
115 ± 10% V, 60 Hz.			

pipetting/vortexing, incubation, separation, and counting. The operation of each of these stations is described in Section II. B. Assay tubes are moved in racks along a track from station to station.

The instrument is equipped with six positive displacement piston type pumps driven by pipettes whose stroke is regulated by sinusoidal variators. Sinusoidal variators make precision pipetting possible by slowing the piston stroke at the beginning and end of each liquid delivery. In this manner the liquid column "shocks" which are caused by sudden starting and stopping of liquid flow, and which are an important source of pipetting error, are eliminated.

The CONCEPT 4® instrument is interfaced with the Hewlett Packard® 9815A Programmable Calculator.

Overall operation of the system is under microprocessor control.

2. Reagents — General

In order best to discuss the functioning of CONCEPT 4®, it is desirable first to describe the general makeup of the AUTOPAK® chemistries specifically designed for use on the system and the preparation of these chemistries prior to use.

A typical reagent package is shown in Figure 2. In this package all components necessary to run the assay are included, except for water and patient sample: standards, controls, buffer, tracer, and antibody-treated assay tubes.

The 8 × 50 mm polypropylene assay tubes come ready for use in 10-place plastic racks. The racks are loaded directly onto the CONCEPT 4® instrument.

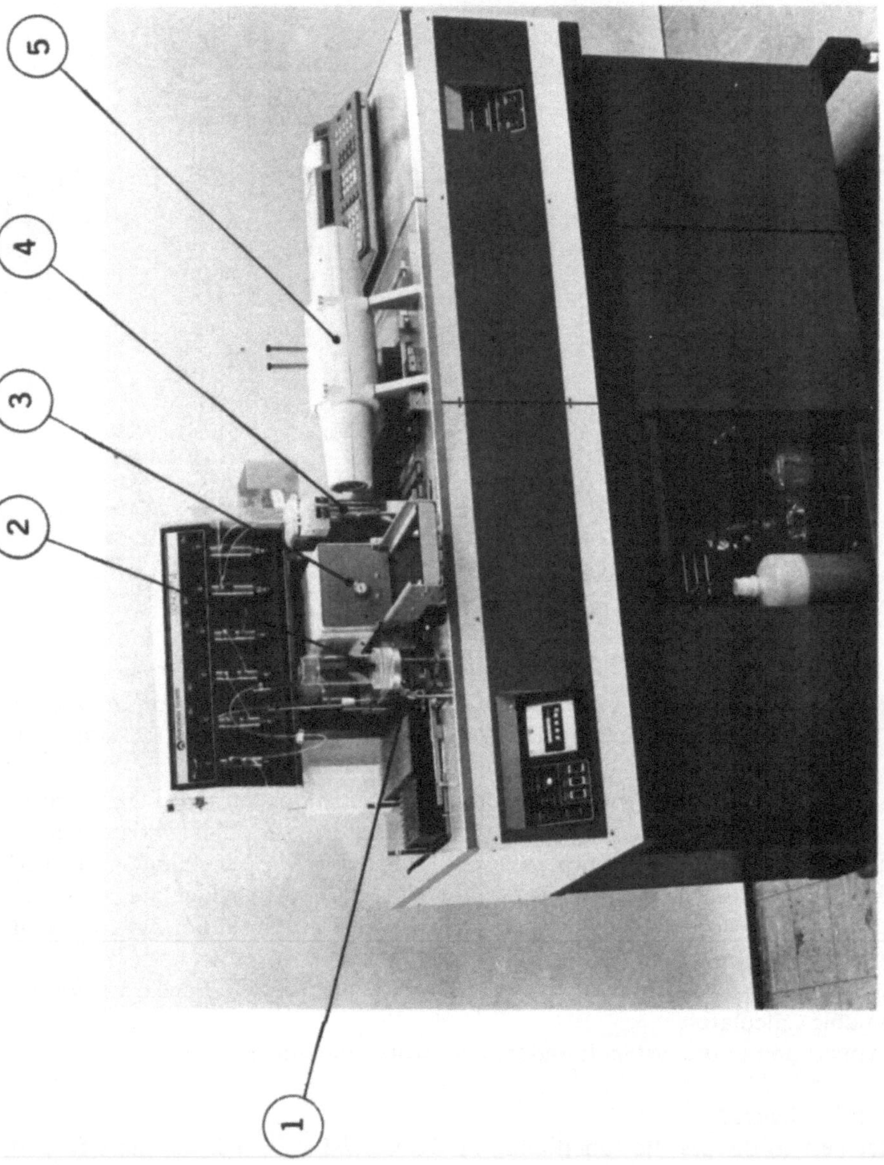

FIGURE 1. CONCEPT 4®.

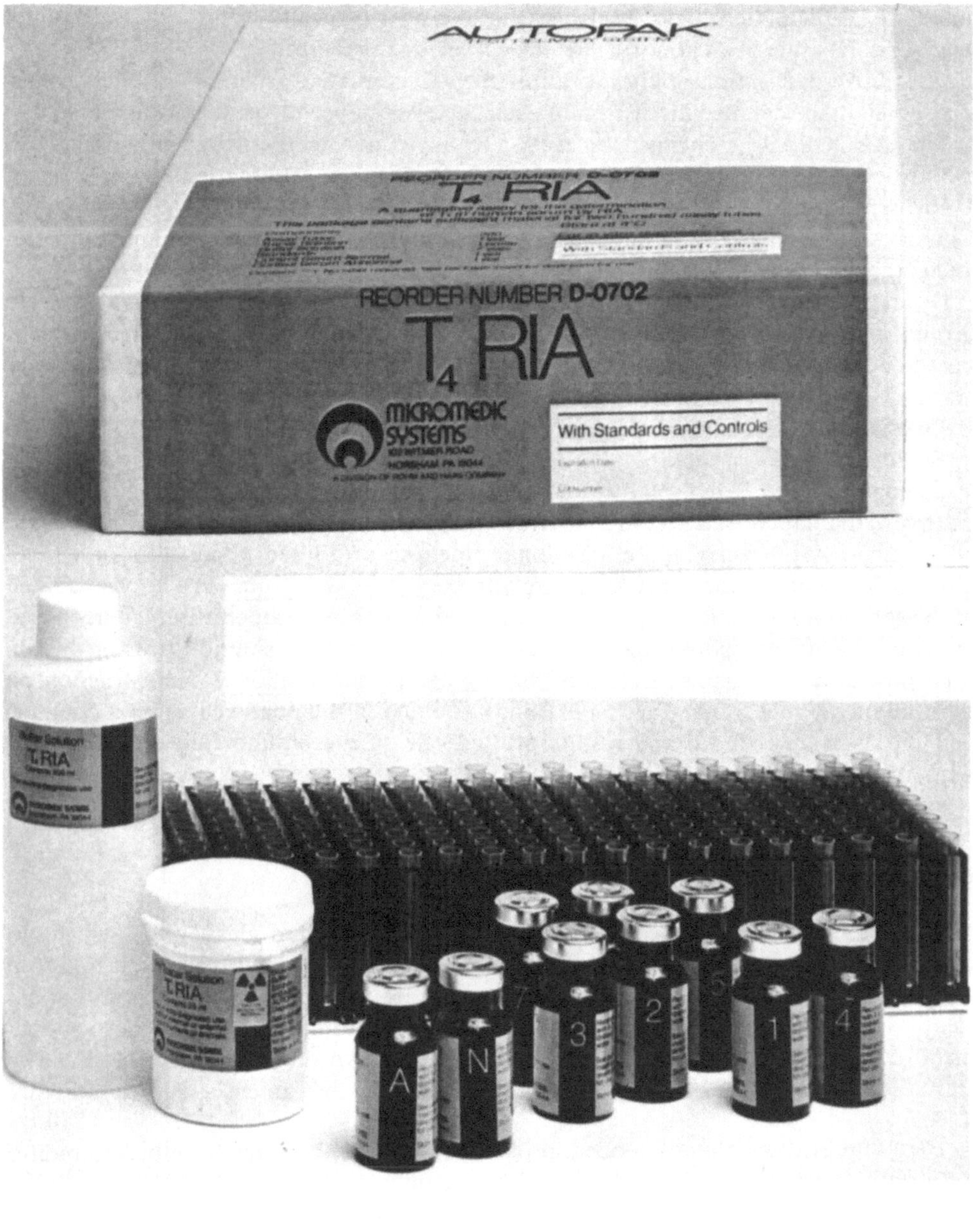

FIGURE 2. AUTOPAK® package and contents.

The tracer generally comes as a liquid in a serum vial, which in turn has been placed in a securetainer to guard against breakage and spillage of radioactive material. The buffer comes in solution form in a polyethylene bottle. The single, diluted tracer reagent solution used in the assay is made up for use simply by mixing together the entire contents of the tracer vial and buffer bottle in a suitable glass container, such as the AUTOPAK® Reagent Dispenser. Excess diluted tracer solution is stored at 4°C in a glass container.

The standards and controls are supplied as lyophilized defibrinated human plasma in serum vials. These are reconstituted for use simply by the addition of 5.0 mℓ of distilled water to each vial. For most assays seven standards, beginning with a "zero" standard, are supplied. Two controls, the one at a clinically normal or therapeutic level, and the other at an abnormal or toxic level, are provided.

The above description applies to all of the AUTOPAK® chemistries unless otherwise noted in the detailed discussion of each chemistry given in Section III below.

The AUTOPAK® chemistries may also be conveniently run using other Micromedic® pipetting and counting equipment, such as the Automatic Pipette, Reagent Dispenser, Automated Pipetting Station, and the 4/200, 2/200, and 4/600 Automatic Gamma Counters. The pipetting may even be done using high quality manual pipettes and competent pipetting techniques.

Depending upon the user's needs, the AUTOPAK® chemistries are also supplied in bulk package form consisting of any desired combination of assay tubes, tracer, buffer, controls, and sets of standards.

B. Operation

1. Summary

The operation of CONCEPT 4® is shown in schematic form in Figure 3. Standards, controls, and patient samples are pipetted from sample tubes into antibody pretreated assay tubes. After receiving an additional pipetting of diluted tracer solution (tracer plus buffer solution) the assay tubes are vortexed. The assay tubes are next incubated in the solid-state incubator for a preset time and at a preset temperature. Upon emerging from the incubator, the free, unreacted sample and tracer solution is washed from each tube at a wash/aspirate station, leaving behind the antibody-bound fraction on the wall of the tube. This fraction is finally counted in the dual-well gamma counter, and the data are reduced and results printed out on the on-line Hewlett Packard® 9815A Programmable Calculator.

The operation of CONCEPT 4® is described in detail in the following sections.

2. Detailed Operation

a. Loading of Sample and Assay Tubes

Samples, standards, and controls are placed in 8 × 50 mm glass tubes, which are placed, in turn, into ten-place plastic racks. The tubes are loaded from right to left in the sequence: standards in ascending order starting with zero, controls, and patient samples. The racks, which contain a notch on the bottom, are then slid onto special metal trays and are held in place by a rib running down the middle of the tray. The trays themselves are then placed on the CONCEPT 4® instrument as shown in Figure 4.

Assay tubes come already loaded in plastic racks which in turn have been mounted on plastic trays. The user simply slides racks containing the desired number of assay tubes onto the metal tray which is placed on the instrument (Figure 4).

For most assays the maximum number of assay tubes that can be loaded at one time is 200. For the T_3 Uptake test the maximum load is 100 tubes.

b. Volume Settings

There are generally two settings which need to be made prior to the start of a given assay: the volume of sample to be delivered to an assay tube, and the volume of reagent (diluted tracer) to be mixed with the sample in the assay tubes. These settings are made simply by dialing in the appropriate number in the small counter above each pump (see Figure 5).

The settings of the other four pumps shown in Figure 5 are generally not changed from assay to assay. The functions of these pumps are discussed more fully below.

In cases where very small sample volumes are required (as little as 2 $\mu\ell$), a 20 $\mu\ell$

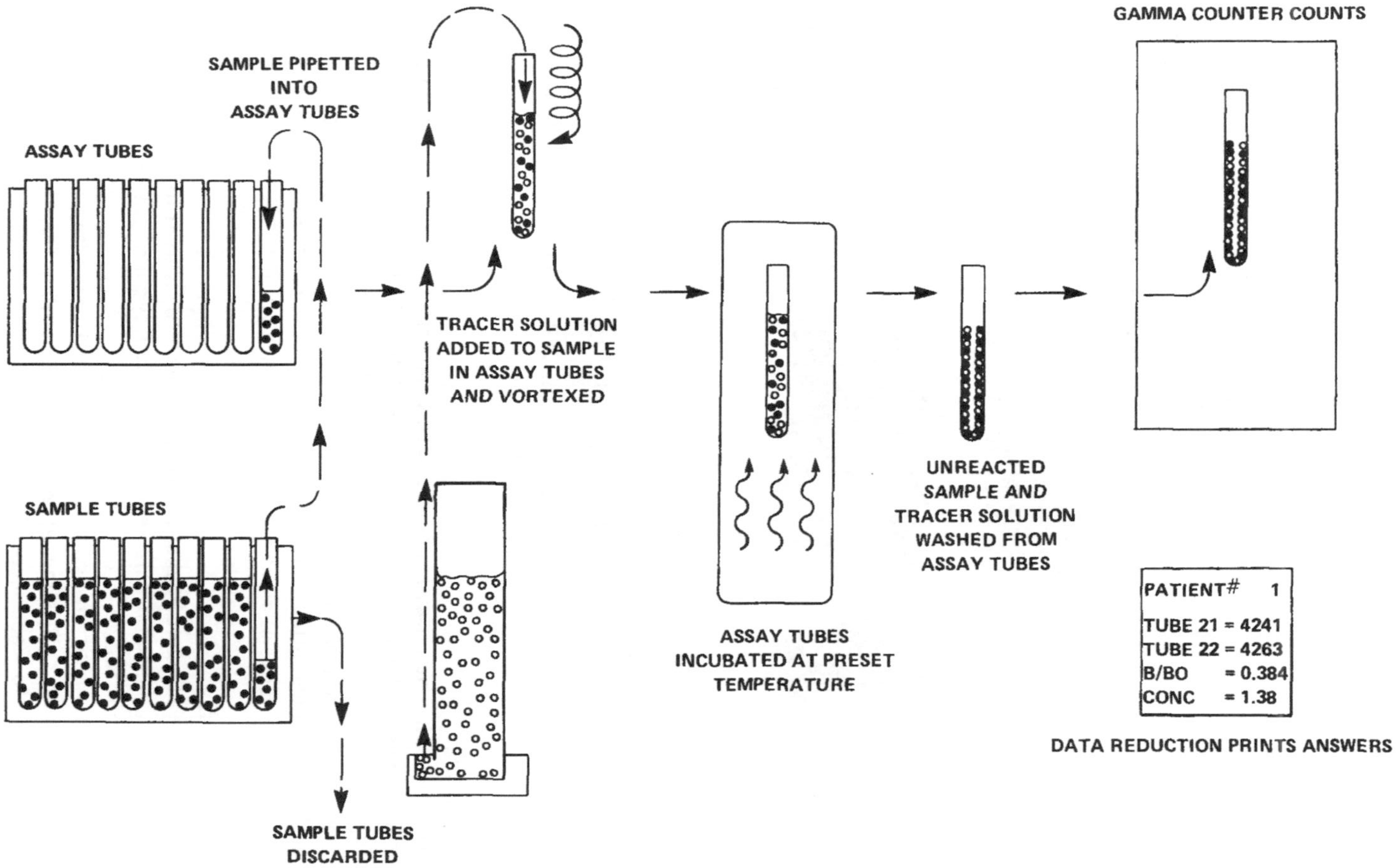

FIGURE 3. Schematic operation of CONCEPT 4®.

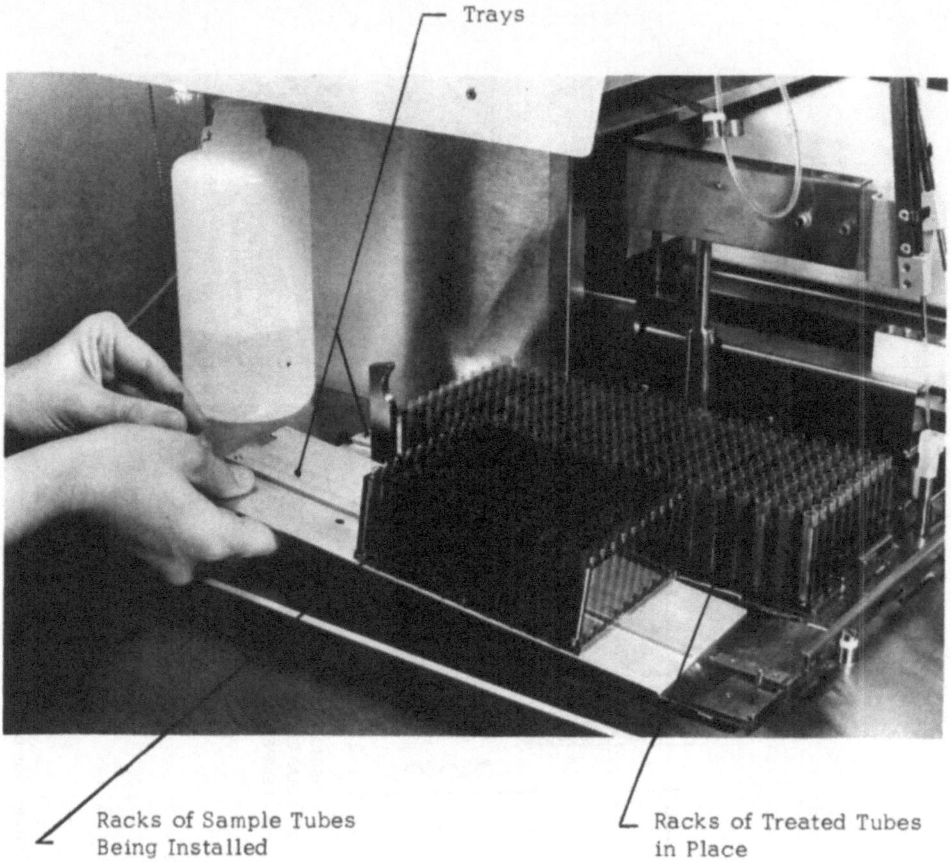

FIGURE 4. Loading trays of racks.

(nominal volume) sample pump is substituted for the 200 $\mu\ell$ pump normally used. A special microsampling tip is also employed under these circumstances.

c. Electronic Settings

Preliminary to beginning an assay, several assay parameters are set using the operating controls shown in Figure 6. The SINGLE/DUPLICATE switch is set for the number of assay replicates desired. A choice of incubation temperature is made: ambient, 37°, or 45°. For nearly all assays the INCUBATION TEMPERATURE switch is left set at 37°. Any incubation time may be selected, ranging from 999.9 min down to 1 min for each of the racks of assay tubes processed in a given assay.

d. Keyboard Assay Entry

The keyboard of the Hewlett Packard® 9815A calculator is used to program in the parameters needed to count and data reduce an assay. Keyboard entries are made in response to the questions shown in the printout in Figure 7. An appropriate keyboard entry is made to signify that an assay is to be counted. An assay batch number is assigned by the program; this is merely an accession number. The batch number is incremented by one unit for each successive assay run.

The type of assay being run is indicated by pressing one of eight keys. Five of these keys are dedicated to the following assays: T_4, T_3, T_3 Uptake, Cortisol, and Digoxin. The remaining three keys are dedicated in accordance with the user's requirements.

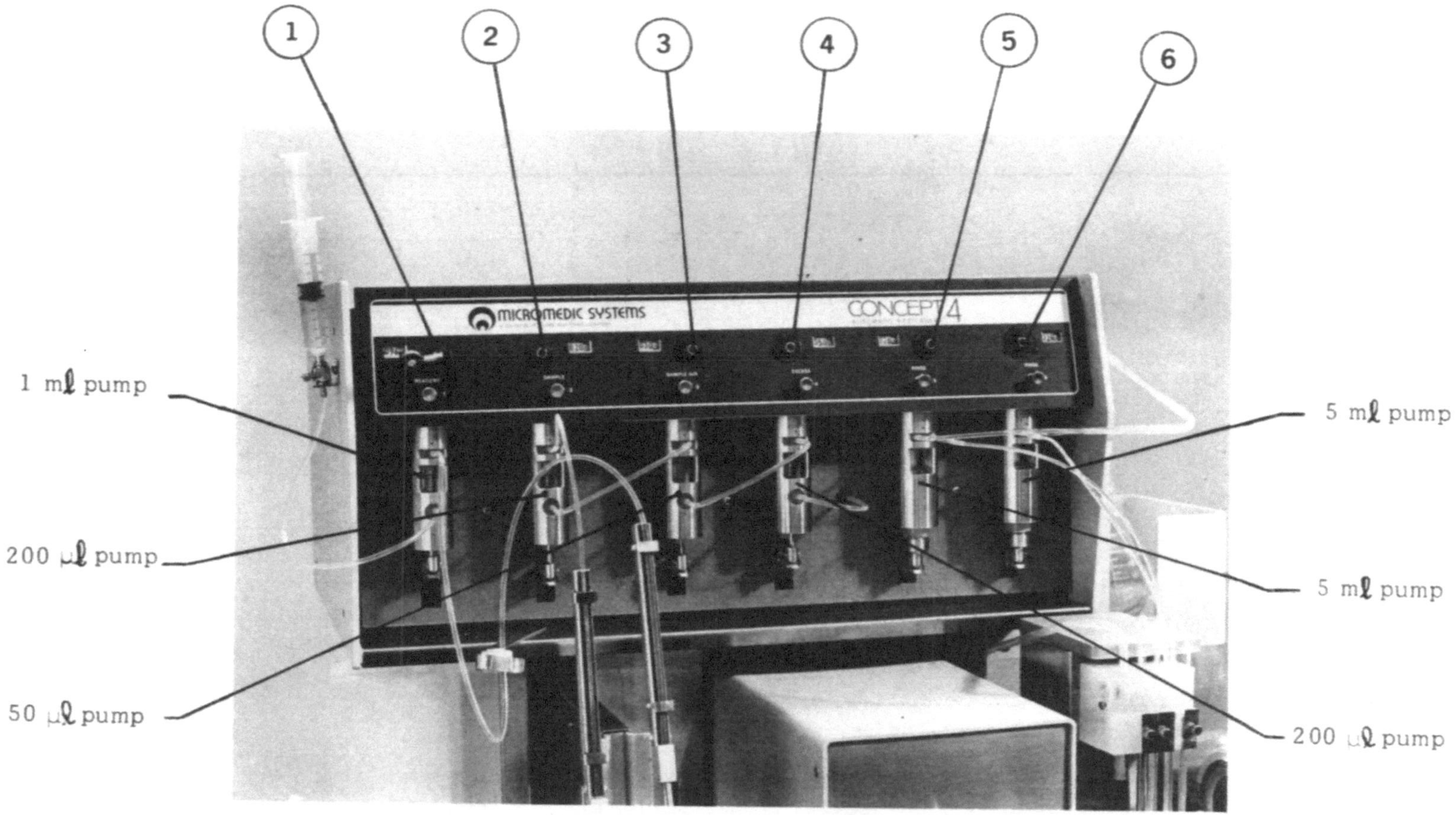

Figure 5. Pipette console. Volume Setting Points: 1. Reagent volume; 2. Sample volume; 3. Sample air volume; 4. Excess sample volume; 5. First wash volume; 6. Second wash volume.

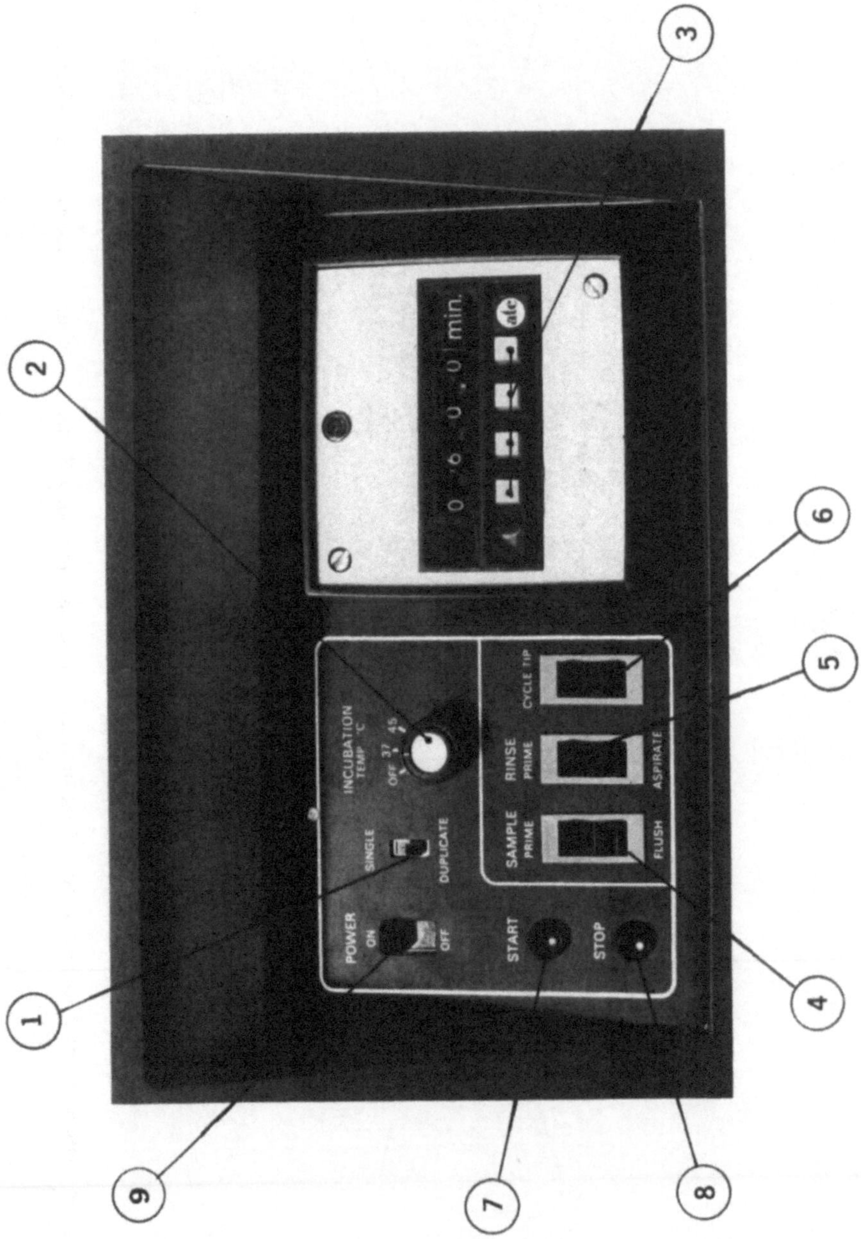

FIGURE 6. Operating controls. 1. Single/duplicate switch; 2. Incubation temperature switch; 3. Timer switches; 4. Sample switch; 5. Rinse switch; 6. Cycle tip switch; 7. Start button; 8. Stop button; 9. Power switch.

```
AUTO START

CONCEPT 4
RADIOASSAY

= = = = = = = = = =

SELECT KEY B-E,

 ENTER ASSAY

BATCH #:          474

ASSAY TYPE?
 (KEYS F-O)
                  T4*
# RACKS?
                   2*
LEFT BACKGROUND
                  35*
RIGHT BACKGROUND
                  28*
COMMENT #?
                   1*
COUNT TIME?
                 1.0*

DATA STORED FOR
BATCH #          474.
```

FIGURE 7. Entry of assay into program.

The pressing of a dedicated key automatically supplies certain information which will be used in data reduction: assay name and number and values of the standards. It also, via the microprocessor, automatically sets the windows of the gamma counter to count the isotope appropriate to the assay.

The "# Racks" question pertains to the number of racks of assay tubes (10 tubes each) being processed.

The background, in terms of counts per minute, is keyed in for the left and right detectors individually. The appropriate background will be subtracted from the counts obtained for each assay tube, regardless of the counting time chosen for the assay.

The response to the question "Comment #" may be the date, operator number, etc.

Any count time, in increments of 0.1 min, within a range of 0.1 to 25.0 min, is permissible.

In the event that the user wishes to perform an assay for which a keyboard key has not been dedicated, the user utilizes the "Special Assay" mode. Again, by means of appropriate responses to questions asked on the printout, the assay is set up by indicating the number of racks in the assay, background counts, comment number, count time, whether bound or free fraction is being counted, presence of total count or non-specific binding tubes, number and values of standards used, number of replicates, and isotope to be counted.

Should the user wish only to obtain assay data in terms of counts per minute, he would utilize the "General Assay" mode. In this case he need only enter the following parameters: number of racks, background, comment number, count time, and isotope.

Up to ten assays may be preprogrammed and stored, ready for counting at any one time. Counting of an assay may be interrupted in order to correct parameters in assays that have not already started counting.

e. Start of Assay

After all settings have been made, the processing of an assay is begun by pressing the START button on the control panel. Sample and assay tube racks each start to advance along parallel tracks toward the first station, where the sample pipetting is done. The advancers on the instrument push racks along by means of spring pushers working in ratchet fashion on ribs on the side of the racks. Racks are advanced in end-to-end fashion.

The pressing of the START button activates a cam which controls the motion of sample and reagent pipetting tips. The pumps to which these tips are connected are not activated, however, until racks of tubes have arrived at the appropriate pipetting stations.

f. Sample Pipetting Station

As the first racks containing samples and assay tubes approach the sample pipetting station (sample racks in front, assay tube racks in rear) (Figure 8), a microswitch is closed. This activates the three pumps involved in sampling: Sample, Sample Air, and Excess.

The sampling cycle is begun by the aspiration of a small air bubble (by the Sample Air pump) into the tip. This bubble creates a barrier between the sample to be picked up and the distilled water which fills the entire pipetting manifold (pumps, line, interconnecting tubing, and tip), preventing unwanted dilution of the sample.

During the next step of the cycle, the tip, whose motion is controlled by the cam discussed in Section II., B., 2., e. above, the Teflon tip descends into the sample tube; the amount of sample to be used in the assay is aspirated by the Sample pump. In addition, a small excess volume of sample is aspirated by the Excess pump. This excess "washes" the walls of the tip free of water and conditions the tip, optimizing pipetting accuracy and precision.

As the tip emerges from the sample tube, it touches the side of the tube so as to remove any extra sample which could cause poor precision in the results. Also, the amount of serum in the sample tube is kept small (500 $\mu\ell$ or less) so that sample does not cling to the outside of the tip.

Next, the tip descends into the assay tube, and the measured amount of sample is expelled from the tip by the Sample pump. As the tip emerges, the droplet of sample, which may be too small to fall of its own accord, is "broken off" against the side of the tube so as to insure a precise delivery.

In the final portion of the cycle the tip moves over a drain, where the excess sample and air bubble are expelled. In addition, the tip is flushed with 1 mℓ of distilled water from a solenoid-controlled reservoir. This rinses out any remaining sample and prepares the tip for the next sample.

After each rack of sample tubes has been completely processed, it is shuttled off onto a discard tray. The assay tube racks continue along the track to the reagent pipetting station.

g. Reagent Pipetting Station

Each assay tube reaches the reagent pipetting station (Figure 9) 24 sec (four positions) after it has left the sample pipetting station. The approach of the first rack of assay tubes trips another microswitch, activating the Reagent pump. In the first part of the reagent transfer cycle the cam-driven Teflon reagent transfer tip descends into the well of the AUTOPAK® Reagent Dispenser. After the Reagent pump has aspirated the designated volume of diluted tracer, the tip moves up and out of the dispenser well and down into the assay tube. The diluted tracer is then dispensed into the assay

FIGURE 8. Sample preparation area.

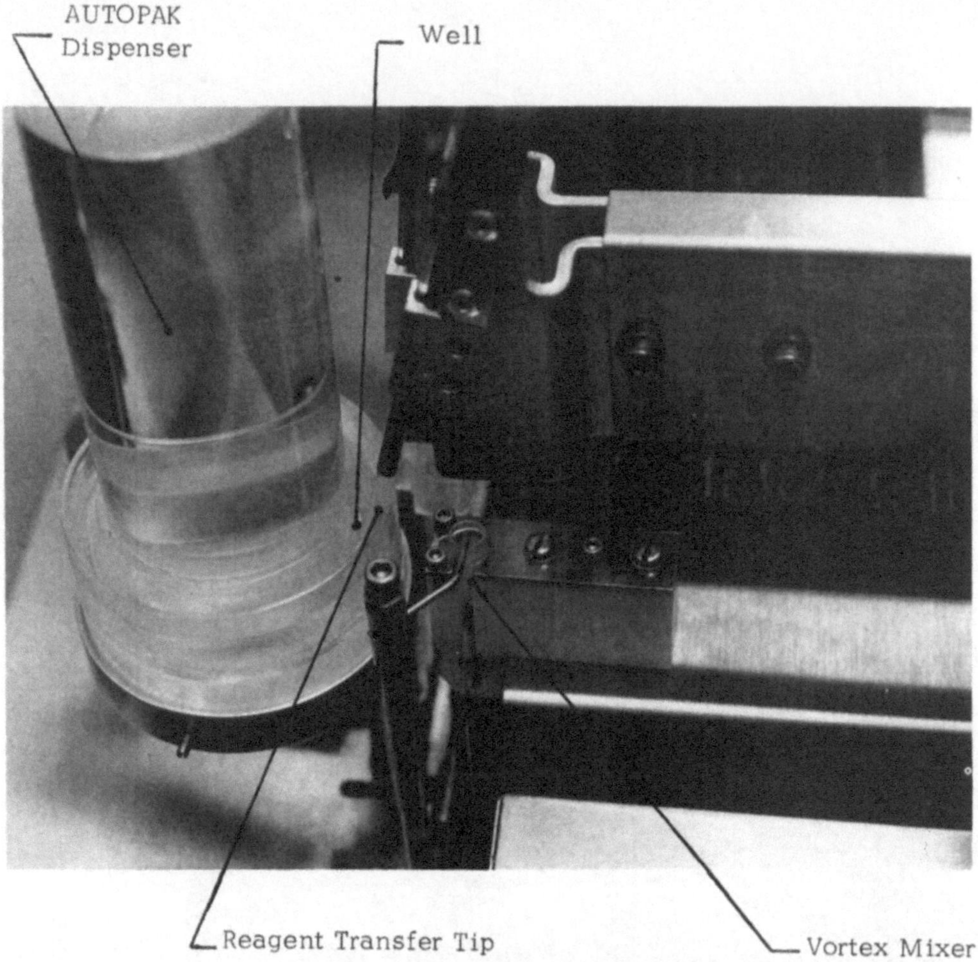

FIGURE 9. Reagent transfer and mixing station.

tube. Finally, the tip emerges from the assay tube, ready to begin a new cycle. The tube, while still in the pipetting station, is mechanically vortexed for about 2 sec before advancing onward with the rest of its rack toward the incubator.

The Reagent Dispenser itself consists simply of a Teflon base (for chemical inertness) with a reservoir well, to which is attached a plain amber glass jar. At the completion of the pipetting for a given assay the dispenser is merely lifted off the instrument, inverted, and the Teflon base removed from the bottle by unscrewing. The bottle, which contains unused reagent, is closed by a screwcap lid and stored at 4°C until the next assay requiring that reagent. The Teflon base, meanwhile, is rinsed and is then ready for the next assay of the day.

h. Incubation

Incubation of assay tubes is performed in a solid state (aluminum alloy block) incubator (Figure 10) which contains sufficient holes to accommodate up to 200 8 × 50 mm assay tubes.

When the first rack of tubes in an assay approaches the incubator, the incubator moves forward one position, and the slide beneath the incubator opens, preparatory to receiving these tubes. When the entire rack of tubes is within the incubator entrance,

FIGURE 10. Incubator.

ten pushrods push the tubes out of the rack and into the first row of holes in the incubator itself. At this point the orange light above the incubator timer comes on, and timing begins.

The incubator next moves forward one more position so as to accommodate the next row of tubes and to keep the first row up in the incubator. The empty first rack is pushed into a storage area beneath the incubator to make way for subsequent oncoming racks.

This loading process is repeated until all of the tubes in the assay have been loaded into the incubator. The slide beneath the incubator then closes, and the incubator moves backward into its "home" position.

When the incubation time expires, the first empty rack in the storage area comes forward into the incubator entrance area; the incubator again moves forward one position; the slide opens; and the first row of 10 tubes is dropped down into this rack. The tubes are pushed down by weighted Teflon balls, one of which is located in each hole of the incubator. The rack then moves out of the incubator area and toward the wash/aspirate station. The remaining rows of tubes are released from the incubator in the same way.

Finally, the incubator slide closes, and the incubator returns to its "home" position, ready for the next assay.

Two important features are built into the incubator to insure optimal precision and accuracy of the assay. One is that uniformity of incubation temperature is maintained over the entire incubator block to within 0.5°C.

The other feature is that assay tubes come out of the incubator in exactly the same order as they went in (first in, first out). This means that all tubes receive exactly the same incubation time. It is therefore possible to significantly shorten the incubation time of many assays, since the assay system need not be at chemical equilibrium in order to obtain good results.

i. Washing/Aspiration

On CONCEPT 4® the separation of the bound fraction from the free fraction in an assay tube is accomplished by means of a wash/aspirate station on the instrument (Figure 11). Since the antibody-bound fraction is in turn bound to the assay tube wall, the free fraction may be readily separated from it by washing.

The wash/aspirate station on CONCEPT 4® consists of three tips: 2 dual wash/aspirate tips and one aspirate tip. When the first rack of tubes in an assay approaches this station from the incubator, a switch is tripped. When the first tube in the rack comes under the first tip, nothing happens. When this tube reaches the position under the second tip, however, the tips come down into the first and second tubes. The front, longer portion of each of these dual wash/aspirate tips aspirates the solution from each tube. The rear, shorter portion of each tip then fills the tubes with approximately 1.4 m*l* of distilled water. Next, the tips emerge from the tubes, and the tubes advance one position. (The dispensing of the aliquots of wash water is made by the two rinse pumps on the right-hand side of the pump manifold shown in Figure 5.) The tips again come down into the tubes. The first tube is aspirated practically dry by the single aspirate tip; the second and third tubes are aspirated and filled with distilled water by the wash/aspirate tips.

The above process continues, with each tube being aspirated and washed twice, and finally aspirated, until the last tube in the assay has been reached. When this tube passes under the second wash/aspirate tip, it is aspirated but not filled with distilled water. When this tube advances under the aspirate tip, the tips do not come down at all.

Note that the first and last tubes in the assay receive only one wash, while all the

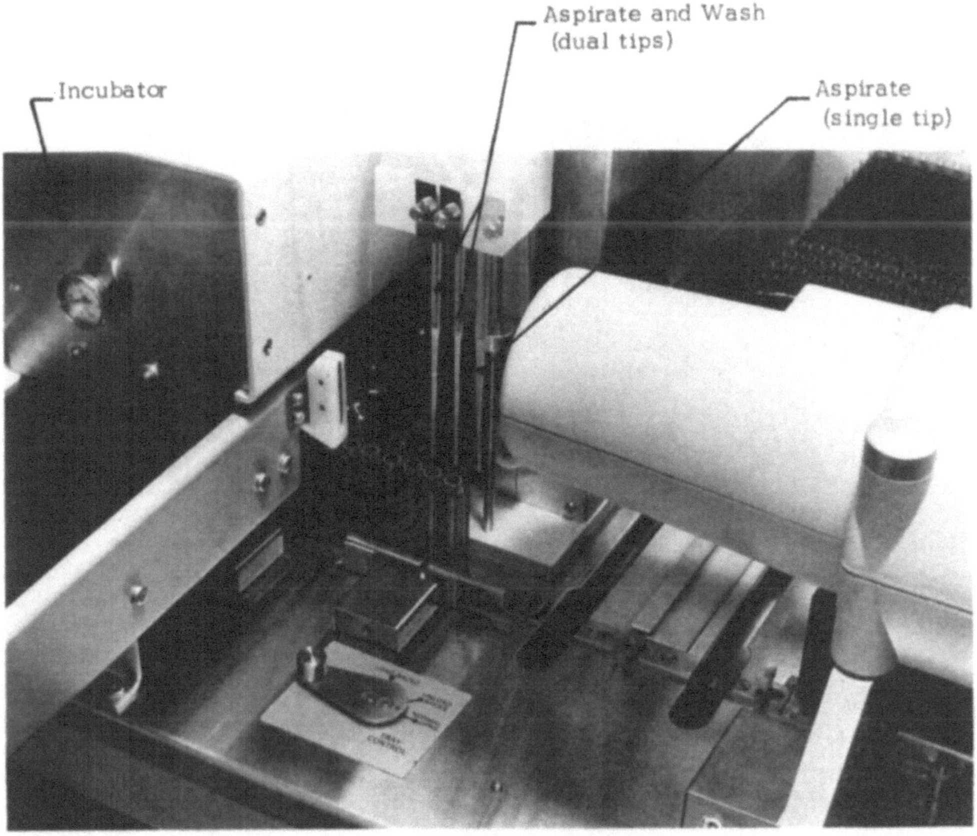

FIGURE 11. Aspirate and wash station.

others have received two. It has been established that one wash is sufficient to remove all unbound tracer from a tube. The single wash condition has been set up for the first and last tube so that water is not spilled onto the top of the instrument by a wash/aspirate tip which has no tube beneath it.

j. Advance into Counter Storage Zone

After leaving the wash/aspirate station, assay tube racks advance further along the track until a microswitch is closed, activating a flipper. The flipper engages the racks, moving them into the precounting storage zone (Figures 12 and 13). This storage area accommodates up to 60 racks (600 tubes) at any one time. (It should also be noted that a similar postcounting storage zone likewise contains up to 600 tubes at a time.)

k. Counting

The gamma counter on CONCEPT 4® is a dual-well counter; the two wells are situated such that the tubes are inserted from the side of the detector rather than from the top (Figure 13). Each detector is equipped with a 2 in. thallium-doped sodium iodide crystal. The crystals are matched to within 1% of each other.

After assay tube racks have advanced along the track into the counting area, the tubes are pushed, two at a time, up into the counter wells by means of two pushrods which come up through holes in the bottom of the racks. The tubes, in turn, push up two weighted rods which project out of the top of the counter.

After each pair of tubes has been counted, the tubes are lowered back into their

FIGURE 12. Counter storage zone.

rack; the weighted rods ensure that the tubes do not stay partially in the counter and cause a jam when the rack advances.

When a rack advances through the counting area, it moves two positions at a time. The first tube to be counted in an assay is always tube 2; it is counted in the left-hand well. The tubes being counted at any one time are three rack positions apart. Tubes are thus counted in the following sequence:

Left well		Right well	
2	and	none;	then
4	and	1;	then
6	and	3;	then
8	and	5;	then
10	and	7;	etc.

Odd-numbered tubes are thus always counted in the right side, and even-numbered tubes are always in the left side.

A printout of the count for a given tube will not appear on the HP9815A® tape immediately upon completion of that count. For example, the count for tube 4 would be stored in memory but would not appear on the tape until tube 3 had been counted. The counts for the two tubes would appear as a pair. In the case of the counting of a standard curve for an assay, no values would appear on the tape until all of the tubes for the entire standard curve had been counted.

The two small digital displays on the control panel beneath the gamma counter (Figure 14) show the counts for the tubes currently being counted or for those most recently completed. When a count exceeds 99,999, an overflow condition occurs, and a red light appears to the left of the numbers in the display.

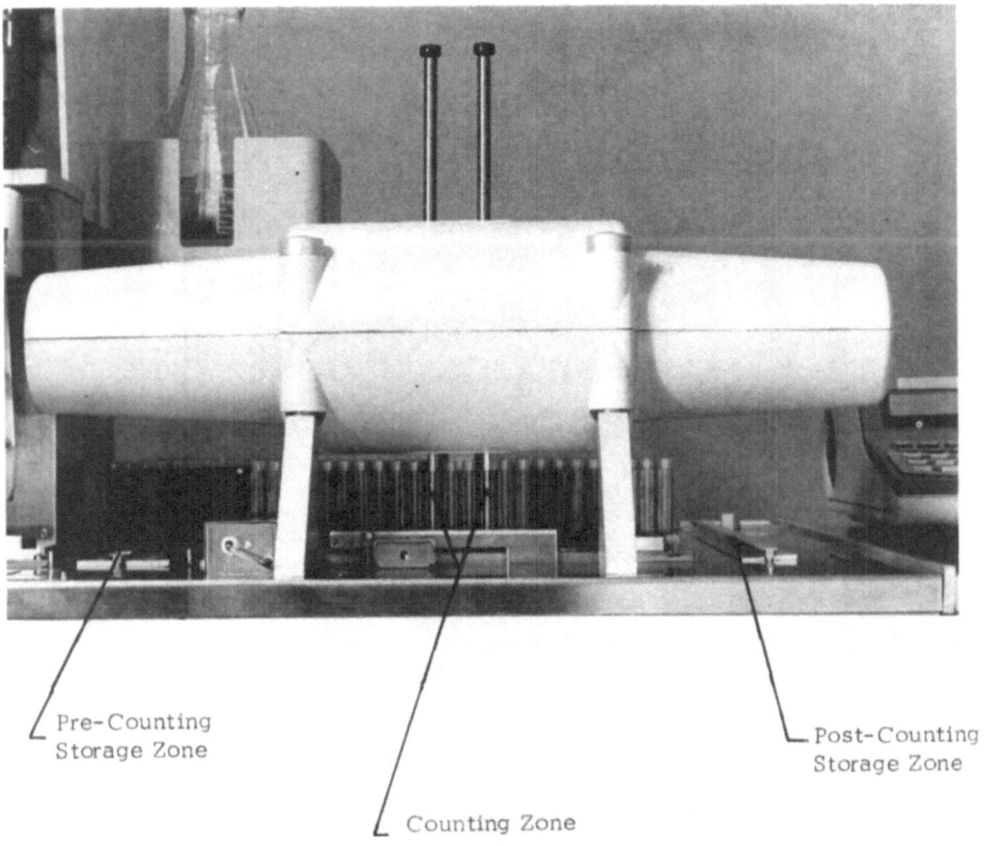

FIGURE 13. Counting station.

1. Data Reduction

Data reduction on the CONCEPT 4® is performed by the Hewlett Packard® 9815A Programmable Calculator, together with the software package devised specifically for CONCEPT 4®. Two basic types of data reduction are performed on CONCEPT 4®: log/logit for radioimmunoassays or proportionality for uptake assays. Additionally, a listing of various quality control parameters may be generated at any time from files of accumulated data. Several types of special data reduction are also available.

The data reduction system is discussed in detail in Section II. C.

3. Special Procedures

In addition to the normal processing, counting, and data reduction of assays, several special operations may be performed using all, or part, of the CONCEPT 4®. These operations enhance the versatility of the instrument with regard to the performance of assays requiring special processing conditions.

a. Sample Predilution

Predilution of a sample may be accomplished by using just the sample transfer and reagent transfer/mixing stations (Figures 8 and 9, respectively) on the instrument. Racks of sample tubes are loaded onto the instrument in the usual fashion. In place of assay tubes are loaded racks containing empty 8 × 50 mm glass tubes. The appropriate diluent is placed in the AUTOPAK® Reagent Dispenser (Figure 9). Racks are

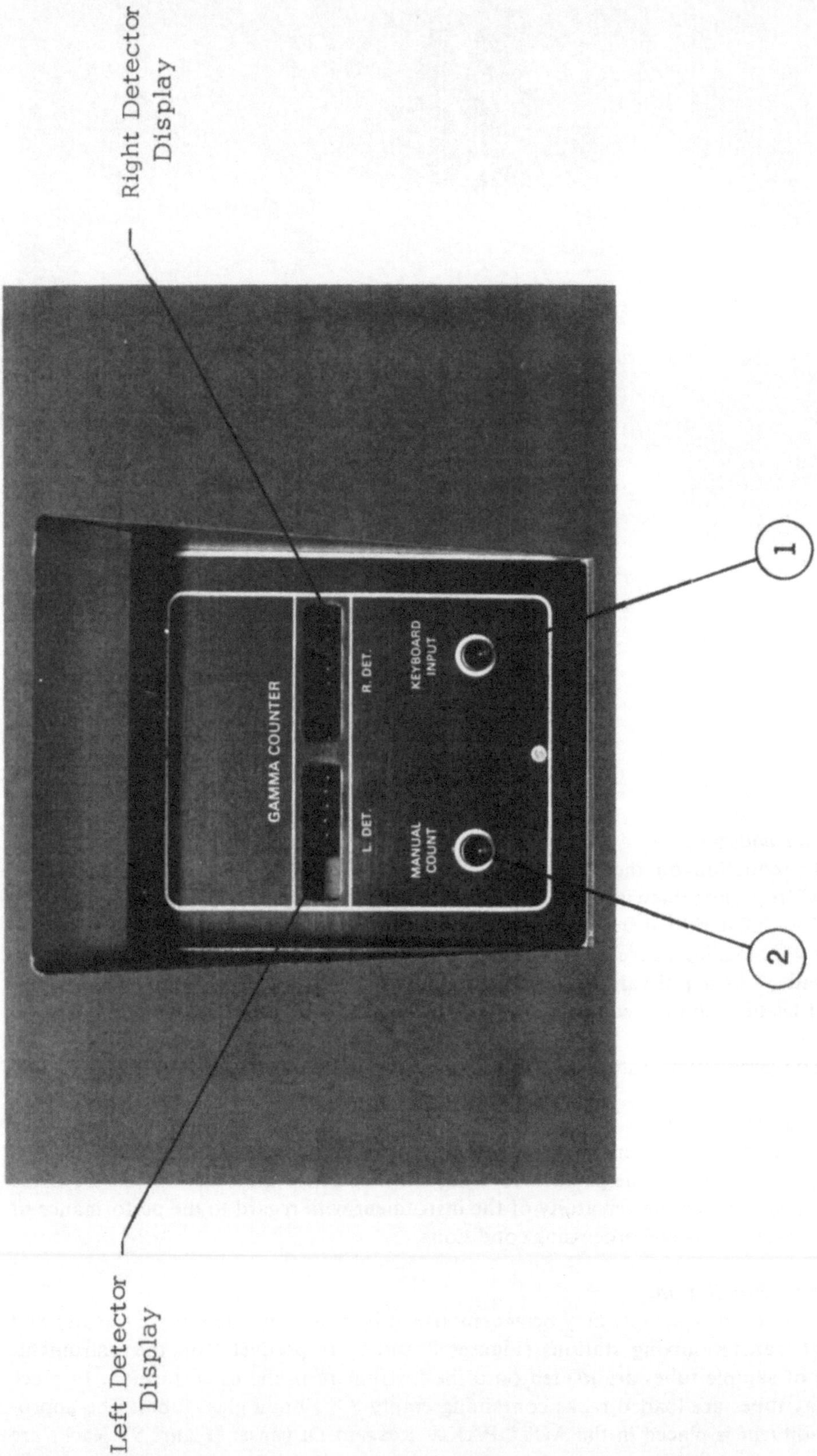

FIGURE 14. Counting station controls.

FIGURE 15. Tray Control selector position for sample dilution.

then processed through the sample transfer and reagent transfer/mixing stations as usual. The incubator timer is left at zero time, so that the racks containing diluted sample will pass directly through the incubator area, rather than be taken up into it. The TRAY CONTROL selection (Figure 15) is set to UNLOAD RACKS so that the wash/aspirate station is not activated and so that the racks will be unloaded rather than pass on into the counter loading zone.

Following this predilution, the diluted sample tubes thus prepared may be placed on CONCEPT 4® as samples, and an assay performed with them in the regular manner.

b. Off-Line Incubation

For assays requiring a relatively lengthy incubation time (or incubation temperatures other than ambient, 37°, or 45°), it is preferable to perform the incubation step off-line, so that subsequent assays may be processed in the interim. In order to accomplish this, assays are set up and processed through the sample transfer and reagent transfer/mixing stations in the normal way. However, the TRAY CONTROL selector is set to the UNLOAD RACKS position as in Section II.B.3.a. above, so that the wash/aspirate station is not activated, and that the racks containing assay tubes may be offloaded near the TRAY CONTROL selector.

At the completion of incubation racks may be replaced on the instrument for further processing in either of two ways:

FIGURE 16. Manual insertion of racks into the counting station.

1. Aspiration and washing of assay tubes is done manually. The TRAY CONTROL selector is set in the LOAD RACKS position (Figure 16), and racks containing washed and aspirated tubes are loaded manually into the counter loading zone. When the TRAY CONTROL selector is returned to the NORMAL OPERATIONS position, the racks are automatically shuttled toward the counting station.
2. Aspiration and washing of assay tubes is done on CONCEPT 4®. The racks containing the incubated assay tubes are loaded onto the instrument in the same position as when beginning an assay. The reagent and sample volumes are set at zero, and the incubator timer setting is left at zero. The TRAY CONTROL selector is in the NORMAL OPERATIONS position. In this way the incubated tubes proceed directly through to the wash/aspirate station, where the free fraction is washed away, and then into the counter loading zone.

In each case, appropriate information is entered on the HP9815A® incubator in the same manner as for a regular assay.

c. Counting Only

When desired, CONCEPT 4® may simply be used as a gamma counter. Racks containing tubes to be counted are loaded manually into the counter loading zone with the TRAY CONTROL selector set in the LOAD RACKS position (Figure 16). In order to initiate counting, the TRAY CONTROL selector is returned to the NORMAL OPERATIONS position and appropriate information is entered into the HP9815A calculator as is done for a regular assay.

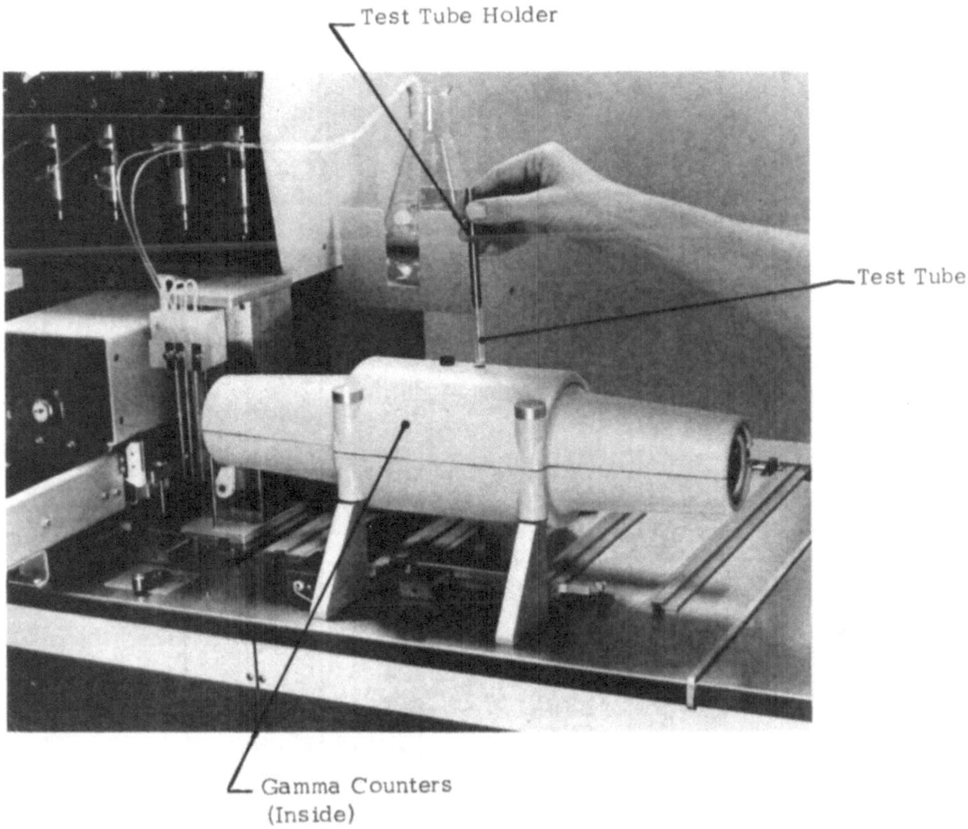

FIGURE 17. Inserting test tube for manual count.

d. Manual Counting

Access may be gained to the Manual Count mode on CONCEPT 4® at any time, even during the counting of another assay, by pressing the KEYBOARD INPUT button on the gamma counter control panel (Figure 14). If tubes are being counted at the time, the counting of these tubes is completed before access to the HP9815A® keyboard is given. After indicating on the keyboard that manual counting is desired, together with the appropriate count time and choice of isotope, the tube to be counted is placed on a special test tube holder and is inserted through the top of either gamma counter detector (Figure 17). Counting is initiated by pressing the MANUAL COUNT button to start each count.

The Manual Count mode is terminated by again pressing the KEYBOARD INPUT button. Counting resumes automatically on the assay, if any, that was being counted before the interruption. The use of the Manual Count mode during the counting of an assay in no way interferes with the counting sequence or data reduction of the assay being processed.

The calibration of the two detectors in the gamma counter may be checked for efficiency, consistency over time, and consistency with each other by means of the Manual Count mode and the simulated, calibrated ^{125}I source supplied with the instrument.

C. Data Reduction System

1. On-Line Data Reduction

a. Reduction of Assay Data — Regular

Examples of the two chief types of assay data reduction performed on CONCEPT

```
T4
BATCH #:            475
COMMENT # :           2
...................
DOSE =             0.00
   TUBE 1 =
                  13937
   TUBE 2 =
                  13580

   BO =           13759
   B/BO =         1.000
. . . . . . . . . . . .
DOSE =             1.00
   TUBE 3 =
                  10879
   TUBE 4 =
                  11592

   B/BO =         0.817
. . . . . . . . . . . .
DOSE =             2.00
   TUBE 5 =
                   9488
   TUBE 6 =
                   9697

   B/BO =         0.697
. . . . . . . . . . . .
DOSE =             4.00
   TUBE 7 =
                   7502
   TUBE 8 =
                   7993

   B/BO =         0.563
. . . . . . . . . . . .
DOSE =             8.00
   TUBE 9 =
                   5452
   TUBE 10 =
                   5543

   B/BO =         0.400
. . . . . . . . . . . .
DOSE =            16.00
   TUBE 11 =
                   3514
   TUBE 12 =
                   3625

DOSE =            32.00
   TUBE 13 =
                   2122
   TUBE 14 =
                   2232

   B/BO =         0.158
....................
R² =              0.994
....................

     WEIGHTED
    REGRESSION
SLOPE =          -2.095
Y INTERCEPT =
                  1.481
ED50 =            5.092
. . . . . . . . . . . .
DOSE =             1.00
CALC. DOSE =       0.98
DIFF. =            2.4%
        . . . .
DOSE =             2.00
CALC. DOSE =       2.04
DIFF. =            1.8%
        . . . .
DOSE =             4.00
CALC. DOSE =       3.85
DIFF. =            3.7%
        . . . .
DOSE =             8.00
CALC. DOSE =       7.97
DIFF. =            0.4%
        . . . .
DOSE =            16.00
CALC. DOSE =      16.13
DIFF. =            0.8%
        . . . .
DOSE =            32.00
CALC. DOSE =      31.98
DIFF. =            0.1%
....................

. . . . . . . . . . . .
CONTROL #1
   TUBE 15 =
          4908
   TUBE 16 =
          4895
   B/BO=0.356
CONC =             9.76

CONTROL #2
   TUBE 17 =
          3749
   TUBE 18 =
          3694
   B/BO=0.270
CONC =            15.15
. . . . . . . . . . . .
CONTROL #3
   TUBE 19 =
          4879
   TUBE 20 =
          5239
   B/BO=0.368
CONC =             9.24
. . . . . . . . . . . .
PATIENT #             1
   TUBE 21 =
          5616
   TUBE 22 =
          5654
   B/BO=0.410
CONC =             7.61
. . . . . . . . . . . .
PATIENT #             2
   TUBE 23 =
          5618
   TUBE 24 =
          5049
   B/BO=0.409
CONC =             7.62
. . . . . . . . . . . .
PATIENT #             3
   TUBE 25 =
          7714
   TUBE 26 =
          7426
   B/BO=0.550
CONC =             4.08
. . . . . . . . . . . .
PATIENT #             4
   TUBE 27 =
          8566
   TUBE 28 =
          8717
   B/BO=0.628
CONC =             2.86

QC STORED
= = = = = = = = = = = =
```

FIGURE 18. Log/logit assay printout.

4®, namely, log/logit and proportionality (T_3 Uptake), are shown in Figures 18 and 19, respectively.

The proportionality calculation performed in reducing T_3 Uptake data is quite simple. Since only one standard is used for this test, the first four assay tubes in a given assay are devoted to this standard. On the HP9815A® printout, the counts per minute of the first four tubes are shown together. Directly underneath this group of counts is shown the average of the first four tubes and the percent uptake of the standard (as given on the vial label and keyed into the HP9815A® when the assay is set up).

In the printout shown in Figure 19, the controls and patient samples have been run in duplicate. For a given sample the counts of the two tubes are averaged and the percent uptake of the sample calculated according to the following equation:

```
T3 UPTAKE

BATCH #        98
COMMENT  # 1477
----------------
  TUBE 1=
          4977
  TUBE 2=
          5062
  TUBE 3=
          5053
  TUBE 4=
          5040
================
 41.6% STANDARD
 5033  CPM AVG.
================
CONTROL  #1
  TUBE 5=
          5470
  TUBE 6=
          5472
UPTAKE=   45.2%
----------------
CONTROL  #2
  TUBE 7=
          4460
  TUBE 8=
          4478
UPTAKE=   37.0%
----------------
CONTROL  #3
  TUBE 9=
          5234
  TUBE 10=
          5257
UPTAKE=   43.4%
----------------
PATIENT  #      1
  TUBE 11=
          5360
  TUBE 12=
          5409
UPTAKE=   44.5%
PATIENT  #      2
  TUBE 13=
          5931
  TUBE 14=
          5963
UPTAKE=   49.2%
----------------
PATIENT  #      3
  TUBE 15=
          4432
  TUBE 16=
          4590
UPTAKE=   37.3%
----------------
PATIENT  #      4
  TUBE 17=
          5196
  TUBE 18=
          5254
UPTAKE=   43.2%
----------------
PATIENT  #      5
  TUBE 19=
          5257
  TUBE 20=
          5041
UPTAKE=   42.6%
----------------

QC  STORED.
```

FIGURE 19. T_3 uptake assay printout.

$$\frac{\%\text{ uptake unknown}}{\%\text{ uptake standard}} = \frac{\text{Average cpm unknown}}{\text{Average cpm standard}} \tag{2.1}$$

The resulting uptake value is shown on the bottom line of the printout for a given control or patient.

The data reduction program for radioimmunoassay generally utilizes seven standards, assayed in duplicate, beginning with zero and proceeding in ascending order, as in the typical T_4 assay shown in Figure 18.

For each standard the following information is listed: dose (concentration) of that standard, the bound counts per minute for each of the duplicate assay tubes, and B/B_o. The average bound counts per minute of the zero standard (Dose = 0.00) is, of course, B_o.

On the basis of the standard curve data, an unweighted logit/log linear regression line is "drawn" mathematically. The correlation coefficient, R^2, is calculated on the basis of the overall fit of this line to the individual data points.

The next calculation consists of five cycles of iterative weighted regression in order to de-emphasize curve outliers and points with poor replication. All data below the "WEIGHTED REGRESSION" notation pertain to the weighted regression line derived from this calculation. The DOSE/CALC. DOSE/DIFF. section provides a convenient means of assessing the goodness-of-fit of the weighted regression at all points along the standard curve.

b. Free Thyroid Index

Free thyroid index (FTI) values can be generated by the CONCEPT 4® data reduction system from individual T_4 RIA and T_3 Uptake values, using a special program tape.

The free thyroid index is calculated from the following equation:

$$FTI = \frac{T_4 \times T_3 \text{ Uptake}}{F} \tag{2.2}$$

where the T_4 value is in $\mu g/d\ell$, T_3 Uptake is in percent and F is a constant factor whose value is chosen so as to place the FTI values on an appropriate scale.

In order to obtain FTI values the T_3 Uptake assay is run first; the T_4's are run subsequently but need *not* be processed directly after the T_3 Uptake run. Naturally, the samples and controls must be kept in the same sequence for both assays. If the user has, for example, some samples for which a T_4 but not a T_3 Uptake has been requested, these samples would be placed at the end of the T_4 run. It is not necessary that both assays be run in singlicate or both in duplicate. For example, the T_4 assay may be run in duplicate and the T_3 Uptake in singlicate.

c. Neonatal T_4 Screening

The data obtained for neonatal T_4 screens is treated by either of two methods. The one method simply involves the assignment of a fixed numerical cutoff point (in $\mu g/d\ell$) to the data from a T_4 assay.

The other method, recommended by Dussault (recognizing that the distribution of T_4 values in a large random population of neonates is a log normal one) is to calculate the geometric mean of a large number of assay values and to establish the cutoff point as being that value which falls 2.0 SD (based on the log normal distribution) below the geometric mean.[1]

The CONCEPT 4® data reduction system may be specially programmed to establish a cutoff point (by preassignment or by calculation) and to "flag" low values by either of the above methods. With either program the assay data are first reduced as shown in Figure 18. Next, the cutoff value is calculated (if the second method is used) and listed. Finally, the assay data are recapitulated in the form of Patient Number/Concentration Value. Immediately beneath any concentration falling below the cutoff point appears the notation "BELOW CUTOFF", together with a row of asterisks.

2. Quality Control List

The "QC List" feature of the CONCEPT 4®/HP9815A® program provides the user with an important means of setting up a quality control program to monitor the performance of each of the preprogrammed tests he performs on CONCEPT 4®.

A QC list is generated in the HP9815A® by pressing the key QC LIST, followed by the key for the assay one wishes to monitor. In Figure 20 is shown an example of a QC list for T_3.

The first items given on the QC list are headed, "Batch #". Below this heading are given the batch numbers of the last T_3 assays (up to 20) run on that instrument. All of the information given on the remainder of the QC list is taken from these assays. When the number of T_3 assays exceeds 20, the earliest assay batches, together with their QC information, are dropped from the list in order to accommodate the newest assays.

```
SELECT KEY B-E

     QC LIST

WHICH ASSAYS?
   (KEYS F-J)
T3
. . . . . . . . . . .
BATCH #
- - - - - - - - - - -
         204
         214
         215
         216
         224

. . . . . . . . . . .
SLOPE
- - - - - - - - - - -
       -2.288
       -2.297
       -2.301
       -2.283
       -2.268
MEAN=     -2.287
 +2SD=    -2.261
 -2SD=    -2.313

. . . . . . . . . . .
Y INTERCEPT
- - - - - - - - - - -
         1.231
         1.343
         1.251
         1.312
         1.361
MEAN=      1.300
 +2SD=     1.413
 -2SD=     1.187

. . . . . . . . . . .
ED50
- - - - - - - - - - -
         3.451
         3.444
         2.915
         2.813
         3.191
MEAN=      3.163
 +2SD=     3.752
 -2SD=     2.574

. . . . . . . . . . .
```

```
BO
- - - - - - - - - - -
          8817
          8646
          8054
          8447
          8439
MEAN=          8481
 +2SD=         9051
 -2SD=         7911

. . . . . . . . . . .
CONTROL 1
- - - - - - - - - - -
          1.14
          1.24
          1.15
          1.19
          1.17
MEAN=          1.18
 +2SD=         1.26
 -2SD=         1.10
. . . . . . . . . . .
CONTROL 2
- - - - - - - - - - -
          3.02
          2.89
          2.79
          2.98

MEAN=          2.90
 +2SD=         3.10
 -2SD=         2.70
. . . . . . . . . . .
CONTROL 3
- - - - - - - - - - -
          1.03
          1.05
          1.11
          1.07
          1.08
MEAN=          1.07
 +2SD=         1.13
 -2SD=         1.01
```

FIGURE 20. Quality control list.

The next two parameters on the QC list are the "Slope" and "Y-Intercept". For a given assay, these two numbers define the logit slope and logit intercept of the logit/log regression line generated by the standard curve data.

The next parameter on the QC list, the "ED_{50}", indicates the T_3 concentration calculated from the point on the standard curve at which $B/B_o = 50\%$ (i.e., the 50% Bound point).

The next item, B_o, merely shows the counts per minute obtained for the "zero-dose" standard for each of the batches in the list. Since the number of counts obtained depends heavily upon the age of tracer being used for a given batch, considerable variation is to be expected in this parameter.

The final three parameters given are "Control 1," "Control 2," and "Control 3." The "Control 1" and "Control 2" values show the actual concentrations obtained, in each of the assay batches listed, on the AUTOPAK®-supplied Normal and Abnormal controls, respectively. "Control 3" is used for an in-house control of the user's choice.

As a further aid in assessing test performance, the user is provided with "Mean", "+2SD", and "−2SD" values at the end of the listing of individual numbers for each of the parameters discussed above. These values, of course, lend themselves readily to the development of Levy-Jennings type wall charts.

In the event that the user ruins an assay by, for example, using the wrong assay parameters, by omitting reagent addition, or by a simple tube mixup, the data file used to generate the QC list may be edited so as to eliminate data from the "bad" assay. In this manner the user is able to avoid the distortion of his mean and standard deviation values that would result from the inclusion of such data.

3. Off-Line Data Reduction

Several options are available for those users who wish to reduce or store data off-line. One option is a direct hardwire interface to, for example, a central computer. Another option is a paper tape punch. In this case, the tape containing the encoded data can, of course, be fed into an appropriate reader which is interfaced with the desired calculator/computer or data storage device.

Two data output formats are available for either of the options mentioned above. The one format lists raw count data in tube number/counts per minute form. The other format provides patient number and concentration data for each sample.

III. AUTOPAK TEST DELIVERY SYSTEM CHEMISTRIES

A. T_4 RIA

The assay protocol for the AUTOPAK® T_4 RIA test is shown in the flow chart in Figure 21. The test may be run either on a ten-fold dilution of serum, as shown (the dilution may be conveniently prepared using the Micromedic® Automatic Pipette supplied with the CONCEPT 4® system or on CONCEPT 4® itself) or directly on 5 μl of serum.

Representative precision data are given in Table 2.

The standards supplied with the reagents are 0, 1.0, 2.0, 4.0, 8.0, 16.0, and 32.0 $\mu g/dl$. The sensitivity of this test is 0.1 $\mu g/dl$, based on 95% B/B_o.

The pH 8.6 assay buffer contains ANS and sodium barbital, as well as 0.1% sodium azide as a preservative. The tracer is supplied in barbital buffer.

B. Neonatal T_4

The assay protocol for the neonatal T_4 test is shown in the flow chart in Figure 22. This test is performed using the reagents, standards, and controls of the AUTOPAK® T_4 RIA assay. The only reagent modification is that the diluted tracer solution is prepared by mixing the assay buffer with tracer in equal volumes rather than the usual one volume of tracer to eight volumes of buffer used in the regular assay. The final amount of tracer added to assay tubes for the neonatal test is slightly less than half that for the regular assay.

The patient samples consist of a 1/8 in. disc punched from filter paper blood spots.

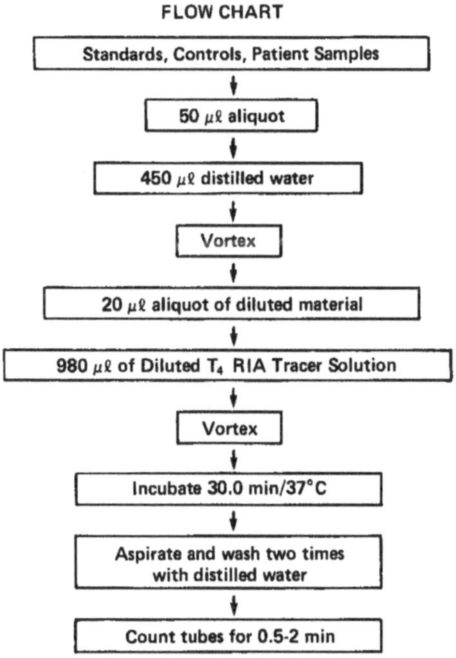

FIGURE 21. AUTOPAK® T_4 RIA flow chart.

Table 2
AUTOPAK® T_4 RIA PRECISION DATA[a,5]

Intra-assay precision (12 replicates/assay)

	Pool 1	Pool 2	Pool 3
$\bar{X}$ (μg/dℓ)	2.37	9.52	19.19
σ	0.10	0.25	0.38
CV%	4.0%	2.6%	2.0%
n	12.0	12.0	12.0

Inter-assay precision (120 replicates, 3 per assay, 40 assays)

	Pool 1	Pool 2	Pool 3
$\bar{X}$ (μg/dℓ)	2.61	9.64	18.28
σ	0.25	0.55	1.21
CV%	9.6%	5.7%	6.6%
n	120.0	120.0	120.0

Inter-assay precision (40 replicates, 1 per assay, 40 assays)

	Normal control	Abnormal control
$\bar{X}$ (μg/dℓ)	8.78	14.70
σ	0.60	0.82
CV%	6.8%	5.6%

[a] Data are taken from assays employing a 60 min incubation time.

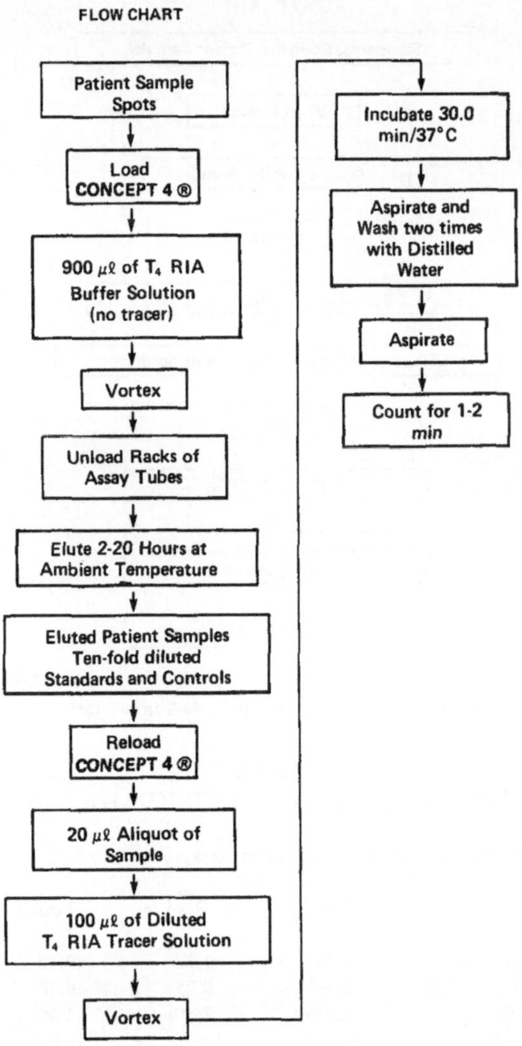

FIGURE 22. Neonatal T_4 assay flow chart.

An in-house blood spot control is also run, in addition to the controls supplied with the reagents.

From the flow chart in Figure 22, it can be seen that a wide-ranging choice of elution time exists. However, once a given elution time has been chosen, that time must be used for all assays to insure reproducible results.

Treatment of the data obtained from the neonatal T_4 assay is discussed in Section II.,C.,1.,C.

C. T_3 RIA

The protocol employed for the AUTOPAK® T_3 RIA assay is shown in the flow chart in Figure 23. Representative precision data are given in Table 3.

The series of standards supplied have values of 0, 0.5, 1.0, 1.5, 2.0, 4.0, and 6.0 ng/mℓ. The sensitivity of this test is 0.25 ng/mℓ, based on $B/B_o = 0.95$.

The pH 8.6 buffer used for the assay contains ANS and sodium barbital, with 0.01% sodium azide as a preservative. The tracer is supplied in a sodium barbital buffer solution.

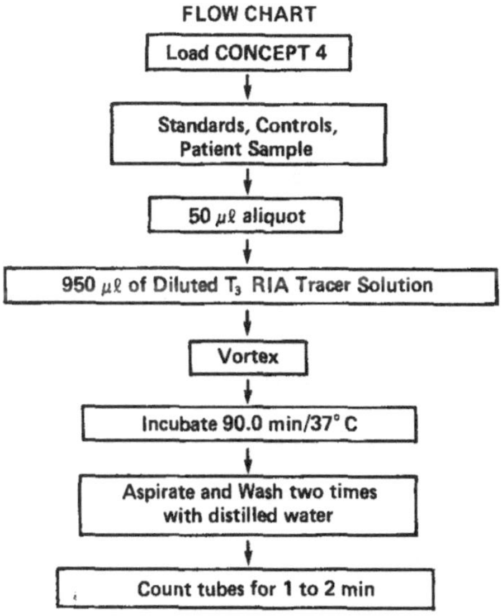

FIGURE 23. AUTOPAK® T_3 RIA flow chart.

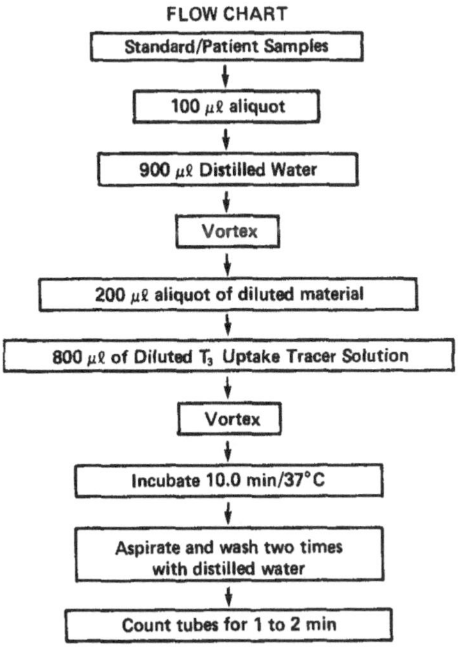

FIGURE 24. AUTOPAK® T_3 uptake flow chart.

D. T_3 Uptake

The protocol used for the AUTOPAK® T_3 Uptake test is shown in the flow chart in Figure 24. The test may be run either on a ten-fold dilution of serum, as shown (the dilution may be made conveniently using either the Micromedic® Automatic Pipette supplied with the CONCEPT 4® system or on CONCEPT 4® itself) or directly on 20 $\mu\ell$ of serum. On samples on which both T_4 RIA and T_3 Uptake tests will be run, it is often preferable to run both tests on a single dilution, rather than directly, since the dilution procedure results in conservation of possibly limited amounts of available sample.

Representative precision data are given in Table 4.

The single standard supplied comes with an assayed, stated value on the label.

The assay buffer contains sodium barbital, at pH 8.0, with 0.1% sodium azide as a preservative. The tracer is also supplied in a sodium barbital buffer solution.

The identical treatment of each assay tube during processing by CONCEPT 4® makes possible the greatly shortened incubation time for this nonequilibrium assay.

In this test the T_3 antibody-treated tube serves merely as a "scavenger" for radioactive T_3 which has not been bound to the sample TBG. Therefore, what is counted in this assay is the *unbound* tracer. In this respect the T_3 Uptake AUTOPAK® assay is entirely analogous to "resin uptake" methodologies.

E. Cortisol

The assay protocol for the AUTOPAK® Cortisol test is shown in the flow chart in Figure 25. This is a direct test for serum or plasma; no sample pretreatment (i.e., protein denaturation) is necessary.

Representative precision data for this assay are given in Table 5.

The standards supplied have values of 0, 2.0, 5.0, 10.0, 20.0, 40.0, and 60.0 $\mu g/d\ell$. The sensitivity of this test is 0.1 $\mu g/d\ell$, based on 95% B/B_o.

Table 3
AUTOPAK® T_3 RIA PRECISION DATA[5]

Intra assay precision (10 replicates/assay)

	Pool 1	Pool 2	Pool 3
$\bar{X}$ (ng/ml)	0.71	1.36	4.05
σ	0.07	0.08	0.15
CV%	9.5%	5.5%	3.8%
n	10.0	10.0	10.0

Inter assay precision (120 replicates, 3 per assay, 40 assays)

	Pool 1	Pool 2	Pool 3
$\bar{X}$ (ng/ml)	0.71	1.20	3.74
σ	0.09	0.13	0.21
CV%	13.9%	11.1%	5.5%
n	120.0	120.0	120.0

Inter assay precision (40 replicates, 1 per assay, 40 assays)

	Normal	Abnormal
$\bar{X}$ (ng/ml)	1.11	2.78
σ	0.09	0.12
CV%	7.8%	4.4%
n	10.0	10.0

Table 4
AUTOPAK® T_3 UPTAKE PRECISION DATA

Intra assay precision (12 replicates/assay)

	Pool 1	Pool 2	Pool 3
$\bar{X}$	31.8	42.4	63.85
σ	0.82	1.14	1.81
CV%	2.6%	2.7%	2.8%
n	12.0	12.0	12.0

Inter assay precision (360 replicates, 9 per assay, 40 assays)
(240 replicates, 6 per assay, 40 assays)

	Pool 1	Pool 2	Pool 3
$\bar{X}$	31.3	41.9	63.3
σ	1.3	1.6	2.3
CV%	4.2%	3.8%	3.6%
n	360.0	240.0	360.0

Inter assay precision (39 replicates, 1 per assay, 39 assays)

	Normal control	Abnormal control
$\bar{X}$	44.32	32.28
σ	1.20	0.97
CV%	2.7%	3.0%
n	39.0	39.0

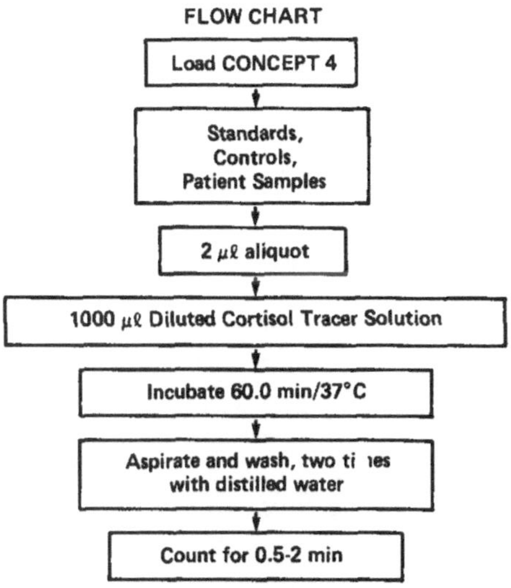

FIGURE 25. AUTOPAK® cortisol flow chart.

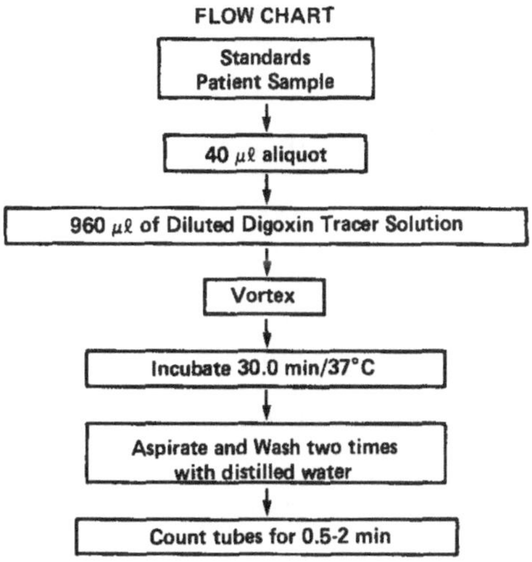

FIGURE 26. AUTOPAK® digoxin flow chart.

The assay buffer used for this test is phosphate buffered saline at pH 7.5; it contains 0.01% sodium azide as a preservative. The tracer is also supplied in phosphate buffered saline.

F. Digoxin

The assay protocol for the AUTOPAK® Digoxin test is shown in the flow chart in Figure 26. Representative precision data are given in Table 6.

The standards provided have values of 0, 0.4, 0.8, 1.6, 2.4, 3.2, and 6.4 μg/mℓ. The sensitivity of this assay is 0.08 μg/mℓ, based on $B/B_o = 0.95$.

The assay buffer is phosphate buffered saline at pH 7.0; the buffer contains 0.01% sodium azide as a preservative. The tracer is also supplied in phosphate buffered saline.

G. Vitamin B_{12}

The procedure for the AUTOPAK® Vitamin B_{12} test is given in the flow chart in Figure 27. The denaturation step in the assay is normally accomplished off-line. Heating to 80°C replaces the boiling step usually required during the denaturation portion of most Vitamin B_{12} assays. Only one incubation is required during the assay proper.

Six standards, at the following levels, are employed: 0, 100.0, 200.0, 400.0, 800.0, and 1600.0 pg/mℓ. They are supplied in defibrinated, denatured plasma, and hence need not be carried through the denaturation portion of the procedure. The controls, on the other hand, are based in defibrinated human plasma and are treated in entirely the same manner as patient samples.

The sensitivity of the AUTOPAK® Vitamin B_{12} assay is 25 pg/mℓ, based on $B/B_o = 0.95$.

A dilute HCl solution is used as denaturant; there is no cyanide in this solution. A small amount of sodium cyanide is contained in the phosphate buffered saline assay buffer (pH 9.5). At no time during the assay is there cyanide in an acid medium.

The tracer is provided in phosphate buffered saline at a basic pH.

The AUTOPAK® Vitamin B_{12} assay is unique in that it is a true radioimmunoassay, rather than a competitive binding protein (intrinsic factor) methodology. Two advantages derive from this fact:

Table 5
AUTOPAK® CORTISOL PRECISION DATA[5]

Intra assay precision (12 replicates/assay)

	Pool 1	Pool 2
$\bar{X}$ (μg/dℓ)	11.43	32.31
σ	0.78	2.47
CV%	7.3%	6.2%
n	12.0	12.0

Inter assay precision (120 replicates, 3 per assay, 40 assays)

	Pool 1	Pool 2
$\bar{X}$ (μg/dℓ)	11.55	35.74
σ	0.98	3.38
CV%	8.6%	9.4%
n	120.0	120.0

Inter assay precision (40 replicates, 1 per assay, 40 assays)

	Normal control	Abnormal control
$\bar{X}$ (μg/dℓ)	11.56	32.19
σ	0.71	1.92
CV%	6.1%	6.0%
n	40.0	40.0

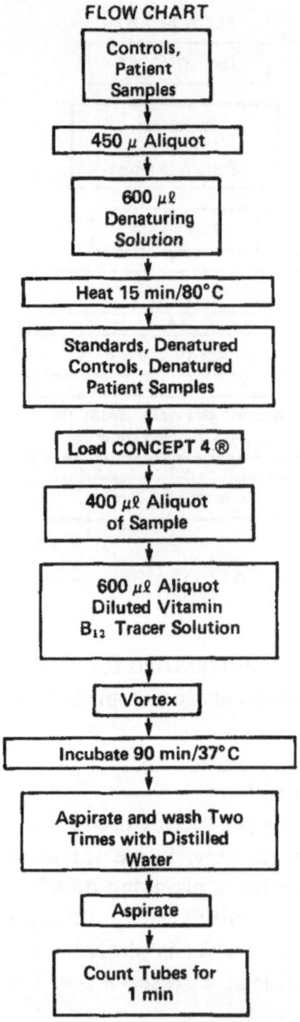

FIGURE 27. AUTOPAK® vitamin B_{12} flow chart.

1. The assay obeys a logit/log relationship, thus greatly simplifying data reduction. Intrinsic factor methodologies do not generally obey this relationship.
2. It has been found that kits employing intrinsic factor as the binding protein yield false normal results in as many as 20% of cobalamin-deficient patients.[2,3] If therapy is withheld or delayed in such patients, severe irreversible neurologic abnormalities may result. The false normals stem from the fact that the binding protein provided in "intrinsic factor" kits incorporate varying large percentages of R proteins. The presence of the R proteins renders the test highly nonspecific by reacting with biologically inactive analogs of Vitamin B_{12}. The AUTOPAK® Vitamin B_{12} test, by virtue of being a true radioimmunoassay, has sharply reduced interference from these analogs.

H. Insulin

The procedure for the AUTOPAK® Insulin assay is shown in the flow chart in Figure 28. Only one incubation step is required for this procedure; this is normally carried out overnight at ambient temperature off CONCEPT 4®.

Table 6
AUTOPAK® DIGOXIN PRECISION DATA

Intra assay precision — CONCEPT 4® (12 replicates/ assay)

	Pool 1	Pool 2	Pool 3
$\overline{X}$ (ng/mℓ)	0.95	1.87	3.51
σ	0.08	0.14	0.23
CV%	8.4%	7.4%	6.5%

Inter assay precision — CONCEPT 4® (120 replicates, 3 per assay, 40 assays)

	Pool 1	Pool 2	Pool 3
X (ng/mℓ)	0.99	2.08	3.89
σ	0.08	0.14	0.23
CV%	7.9%	6.5%	6.0%

Inter assay precision — CONCEPT 4® (40 replicates, 1 per assay, 40 assays)

	Normal	Abnormal
$\overline{X}$ (ng/mℓ)	0.99	2.44
σ	0.08	0.14
CV%	7.6%	5.6%

From Micromedic® Systems clinical validation files. With permission.

Representative precision data for this test are given in Table 7.

The standards provided are porcine insulin lyophilized from phosphate buffered saline containing bovine serum albumin, with preservatives, at the following levels: 0, 10.0, 20.0, 50.0, 100.0, 200.0, and 400.0 μU/mℓ. They are calibrated against the World Health Organization 66/304 International Reference Insulin Standard.

The sensitivity of this test is 8.0 μU/mℓ.

The assay buffer is phosphate buffered saline at pH 7.0, with sodium azide as a preservative.

The tracer is supplied lyophilized from phosphate buffered saline containing bovine serum albumin, with sodium azide as a preservative. It is reconstituted by the addition of 10.0 mℓ of distilled water.

I. hTSH RIA

The assay protocol for the AUTOPAK® hTSH RIA test is shown in the flow chart in Figure 29. Only one, room-temperature incubation step is required for this assay; this is usually carried out overnight off CONCEPT 4®.

The standards provided are porcine serum-based. The following levels are supplied: 0, 2.5, 5.0, 10.0, 20.0, 40.0, and 100.0 μU/mℓ. The sensitivity of this assay is 0.8 μU/mℓ (based on B/B_o = 0.95), equal to or better than the sensitivities obtained with sequential hTSH assays.

The buffer used in this assay is phosphate buffered saline.

The tracer for this assay is supplied in four vials; each contains sufficient tracer for 50 tubes. It is provided in lyophilized form, from phosphate buffered saline which

Table 7
AUTOPAK® INSULIN PRECISION DATA[5]

Intra assay precision[a]

	Pool 1	Pool 2	Pool 3	Pool 4	Pool 5
$\bar{X}$ (μU/mℓ)	39.5	258.6	11.3	83.1	130.4
σ (μU/mℓ)	4.6	12.1	2.6	5.4	7.3
CV (%)	11.6	4.7	22.7	6.5	5.5
n[b]	60.0	60.0	120.0	120.0	120.0
m[c]	10.0	10.0	20.0	20.0	20.0

Inter assay precision[a]

	Pool 1	Pool 2	Pool 3	Pool 4	Pool 5	
$\bar{X}$ (μU/mℓ)	39.5	258.6	11.3	83.1	130.4	
σ (μU/mℓ)	7.2	15.0	3.2	7.6	14.1	
CV (%)	18.2	5.8	28.1	9.2	10.8	
n[b]	60.0	60.0	120.0	120.0	120.0	
m[c]	10.0	10.0	20.0	20.0	20.0	
	Pool A	Pool B	Pool C	Pool D	Pool E	Pool F
$\bar{X}$ (μU/mℓ)	39.4	57.7	11.8	54.8	106.6	144.4
σ (μU/mℓ)	4.2	4.9	2.6	5.2	10.6	10.7
CV (%)	10.7	8.5	21.8	9.5	9.9	7.4
n[b]	30.0	30.0	20.0	20.0	40.0	20.0
m[c]	30.0	30.0	20.0	20.0	40.0	20.0

[a] Calculations were carried out using the Rodbard method.[6]
[b] n = total number of samples.
[c] m = number of assays in which the samples were distributed.

contains bovine serum albumin. Each vial is reconstituted by the addition of 5.0 mℓ of distilled water. All vials contain a small amount of ion-exchange resin, which adsorbs iodine and other molecular fragments resulting from radioactive bombardment; these substances would, if left in solution, reduce the specificity of the assay.

IV. FEATURES/BENEFITS

There are several features of the CONCEPT 4® system from which users derive considerable benefits; these are discussed in the following sections.

A. Performance

The performance of the integrated AUTOPAK®/CONCEPT 4® Automatic Radioassay System is amply attested by the data presented in Section III.

An important factor in the performance of any assay system is consistency. With CONCEPT 4® this consistency is achieved by the total repetition of all steps in an assay, day after day, month after month. Because of this automated, "hands-off" repetition, it is possible to eliminate the errors and variability in assay results associated with technologist-to-technologist performance and technique differences.

The performance of a system is greatly enhanced by the compatibility of the various components. In an integrated instrument/reagent system both major components have been specifically designed for each other. During the development process it is possible to set and adjust the relevant parameters in both components so as to obtain an optim-

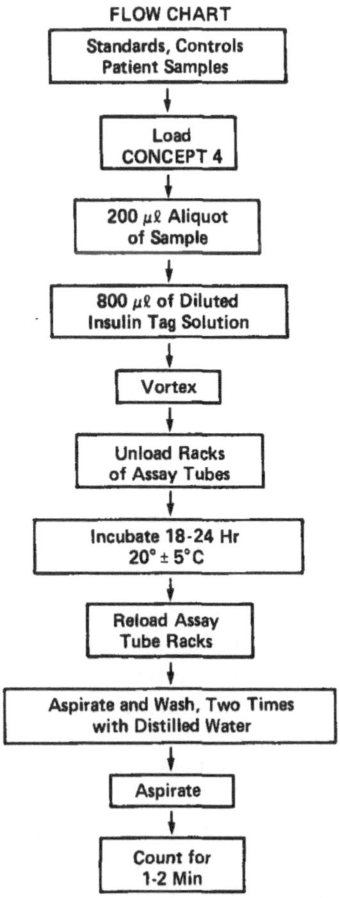

FIGURE 28. AUTOPAK® insulin flow chart.

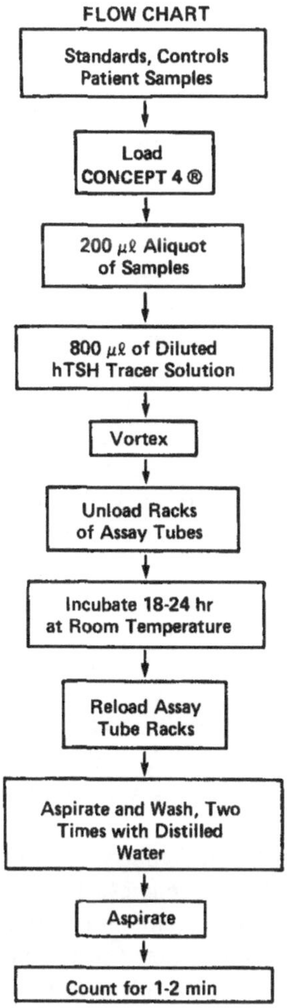

FIGURE 29. AUTOPAK® hTSH flow chart.

ized total system. Frequently, when attempts are made to "mate" a test or reagent from one manufacturer with an instrument from another manufacturer, which has not been designed for that specific test, it is necessary to compromise on parameters such as incubation temperatures and times, sample sizes, and modes of reagent addition; this leads to impaired overall performance. The achievement and maintenance of suitable quality control over the test reagents by the manufacturer is greatly simplified when these reagents will be used on only one instrument system or on several closely related systems. Quite simply, this results from the fact that the manufacturer controls all variables which can affect the assay.

B. Throughput

The development of AUTOPAK® chemistries with a simple, uniform protocol, and the integration of these chemistries with the CONCEPT 4® instrumentation has permitted the development of a highly automated system with very high throughput. The simplification involved in the addition of only a single reagent contributes greatly to the streamlining and speeding up of assay processing. The use of antibody-treated assay tubes eliminates the time-consuming step of centrifugation. Further efficiency is achieved by the elimination of manual manipulation between the various steps of the assay.

The end result of the considerations described above is a system with the speed and capacity to enable laboratories to process substantial workloads within less than a normal eight-hour day, using only 30 min of a technologist's time. The throughput of this system is best illustrated by the throughput analyses given below.

In Figure 30 is shown a time flow chart of assay processing on CONCEPT 4®. The time required for initial setup and loading of the instrument for an assay is approximately 5 min. After the start of an assay proper, about 2 min is required for each rack of tubes to pass through the sample and reagent pipetting station. Incubation requires various times, depending upon the assay in question. After emerging from the incubator, each rack requires 2 min to clear the wash/aspirate (separation) station. Because rack motion takes place at the rate of one rack per minute the additional time required in minutes for the *last* rack of a given assay to clear the wash/aspirate station would be equal to the number of racks in the assay. Therefore, since a subsequent assay can be started as soon as the last rack of the previous assay has cleared the wash/aspirate station, the total required elapsed time between the starting times of any two assays is equal to the sample process time of the previous assay (pipetting + incubation + separation) + the number of racks in the previous assay. The setting up of a subsequent assay may be performed during the processing of the previous assay.

On the basis of the above considerations, throughput analyses and daily schedule charts were developed for several hypothetical laboratories having small, medium, and large workloads; these are shown in Tables 8 through 13.

The following parameters were employed in developing these analyses:

1. It is assumed that five assays are being performed at all three laboratories with the following frequencies: T_3 Uptake and T_4, five days/week each; digoxin, 6 days/week; and cortisol and T_3 RIA, once per week each.
2. The analyses shown are for a typical weekday during which *either* the cortisol *or* the T_3 RIA assay would be run. Logically these two assays would be run on different days, e.g., cortisol on Tuesdays and T_3 RIA on Thursdays.
3. The incubation periods and counting times for the five assays used are those shown in Table 14.
4. All assays are performed in duplicate (standard curve, controls, and unknowns). Seven standards are used for all assays with the exception of T_3 Uptake, together with three controls. The T_3 Uptake assay employs one standard, run in quadruplicate, and three controls.

The total assay time shown for a given run is the sum of the sample process time plus the counting time. The total elapsed time represents the time in minutes from the beginning of the working day (9:00 A. M.).

The time at which a given assay starts to count clearly depends upon the attainment of both of the following conditions: the completion of counting of the previous assay and the completion of sample processing for the current assay. In order to make more efficient use of the gamma counter portion of the instrument, the assays requiring shorter incubation times should therefore be done first during the day. This procedure has the advantage of getting the counter into action as early as possible in the day (9:19 A. M.) and allows counting of earlier assays to proceed while later ones are still at the incubation stage. In addition, results are available for the earlier assays sooner.

With a daily workload consisting of 360 samples (Case III, Table 13), and on a day

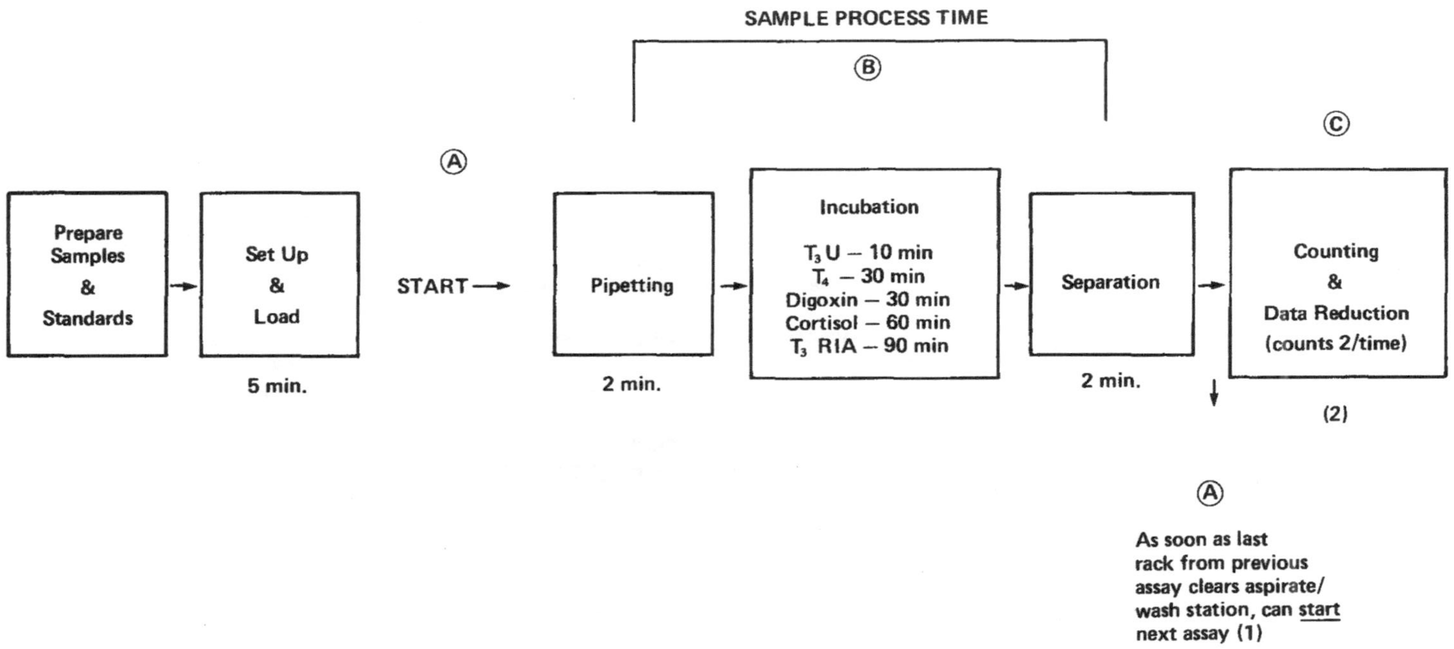

(1) Start next assay = (incubation time + 4) + # of racks in previous assay. (i.e. 1 rack/minute)

(2) Counting Time = $\frac{[\text{Preset time (seconds)} + 2 \text{ seconds}] \times \# \text{tubes}}{120 \text{ seconds}}$

FIGURE 30. Assay process flow chart.

Table 8
THROUGHPUT ANALYSIS FOR CASE I — 55 SAMPLE DAILY WORKLOAD

Test and No. of samples	No. of tubes	Start assay[a]	Sample process time	Start counting[a]	Counting time	Total assay time	Total elapsed time[a]
15 T_3U	40	5 min	14 min	19 min	21 min	35 min	40 min
20 T_4	60	25	34	59	16	50	75
10 Digoxin	40	67	34	101	11	45	112
10 Cortisol	40	107	64	171	11	75	182
or							
10 T_3 RIA	40	107	94	201	21	115	222

[a] Time in minutes elapsed from time 0; i.e., loading first assay.

Table 9
CONCEPT 4® DAILY SCHEDULE FOR: CASE I

Test and no. of samples	No. of tubes	Assay frequency	Start time	Tech time	Total assay time	Results ready
15 T_3U	40	5/week	9:00 a.m.	5 min	35 min	9:40 a.m.
20 T_4	60	5/week	9:25 a.m.	5 min	50 min	10:15 a.m.
10 Digoxin	40	6/week	10:07 a.m.	5 min	45 min	10:52 a.m.
10 Cortisol	40	1/week	10:47 a.m.	5 min	75 min	12:02 p.m.
or						
10 T_3 RIA	40	1/week	10:47 a.m.	5 min	115 min	12:42 p.m.

Table 10
THROUGHPUT ANALYSIS
FOR CASE II — 110 SAMPLE DAILY WORKLOAD

Test and no. of samples	No. of tubes	Start assay[a]	Sample process time	Start counting[a]	Counting time	Total assay time	Total elapsed time[a]
30 T_3U	70	5 min	14 min	19 min	36 min	50 min	55 min
40 T_4	100	28	34	62	27	61	89
20 Digoxin	60	74	34	108	16	50	124
20 Cortisol	60	116	64	180	16	80	196
or							
20 T_3 RIA	60	116	94	210	31	125	241

[a] Time in minutes elapsed from time 0; i.e., loading first assay.

Table 11
CONCEPT 4® DAILY SCHEDULE
FOR: CASE II

Test and no. of samples	No. of tubes	Assay frequency	Start time	Tech time	Total assay time	Results ready
30 T_3U	70	5/week	9:00 a.m.	5 min	50 min	9:55 a.m.
40 T_4	100	5/week	9:28 a.m.	5 min	61 min	10:29 a.m.
20 Digoxin	60	6/week	10:14 a.m.	5 min	50 min	11:02 a.m.
20 Cortisol or	60	1/week	10:56 a.m.	5 min	80 min	12:16 p.m.
20 T_3 RIA	60	1/week	10:56 a.m.	5 min	125 min	1:01 p.m.

Table 12
THROUGHPUT ANALYSIS FOR CASE III — 360 SAMPLE DAILY WORKLOAD

Test and no. of samples	No. of tubes	Start assay[a]	Sample process time	Start counting[a]	Counting time	Total assay time	Total elapsed time[a]
45 T_3U	100	5 min	14 min	19 min	52 min	66 min	71 min
45 T_3U	100	31	14	71	52	66	123
90 T_4	200	57	34	123	53	87	176
90 Digoxin	200	113	34	176	53	87	230
90 Cortisol or	200	169	64	233	53	117	286
90 T_3 RIA	200	169	94	263	103	197	366

[a] Time in minutes elapsed from time 0; i.e., loading first assay.

Table 13
CONCEPT 4® DAILY SCHEDULE FOR: CASE III

Test and no. of samples	No. of tubes	Assay frequency	Start time	Tech time	Total assay time	Results ready
45 T_3U	100	5/week	9:00 a.m.	5 min	66 min	10:11 a.m.
45 T_3U	100	5/week	9:31 a.m.	5 min	66 min	11:03 a.m.
90 T_4	200	5/week	9:57 a.m.	5 min	87 min	11:56 a.m.
90 Digoxin	200	6/week	10:53 a.m.	5 min	87 min	12:50 a.m.
90 Cortisol or	200	1/week	11:49 a.m.	5 min	117 min	1:46 p.m.
90 T_3 RIA	200	1/week	11:49 a.m.	5 min	197 min	3:06 p.m.

Table 14
INCUBATION AND COUNTING TIMES FOR SEVERAL AUTOPAK® ASSAYS

Assay	Incubation (min)	Counting (min)
T_3 Uptake	10	1.0
T_4 RIA	30	0.5
Digoxin	30	0.5
Cortisol	60	0.5
T_3 RIA	90	1.0

in which T_3 RIA, the assay requiring the longest incubation time, is being run, the last patient result of the day is ready by 3:06 P. M. With lighter workloads, all answers are obtained by midday.

C. Time Savings

There are other significant time savings to be had from the use of the CONCEPT 4® system. As mentioned in Section IV., B. above, only 30 min of a technologist's time is required to process large workloads during the course of a normal working day. This is made possible because of the following three features:

1. There is complete "hands off" automation of all operating steps.
2. The reagents are premeasured and prestandardized so that no preparation such as reagent reconstitution or pH adjustment is necessary. Standards and controls are provided in individual vials, each of which is reconstituted with the same 5 mℓ volume of distilled water. No serial dilution is required. The antibody treated tubes come ready to use in special racks, suitable for loading without transfer or other manipulation.
3. Simplicity of changeover from one assay to another is accomplished merely by loading the appropriate assay tubes for the next assay, replacing the reagent dispenser containing the reagent for the previous assay with the dispenser for the next one, and following a brief, simple checklist for sample and reagent volumes, incubation time, etc.

The minimizing of technologist time becomes especially important when one considers the difficulties involved in obtaining adequate staffing in order to handle increasing RIA workloads by conventional labor intensive techniques. There is, first of all, the significant problem of finding people with the training, experience, and motivation suitable for demanding RIA work. Then, of course, there is the cost of maintaining a staff of highly motivated and trained personnel to do the RIA work.

The rapidity of changeover from one assay to another is a key feature in being readily able to run a "stat" digoxin on CONCEPT 4®. It is worth noting that even if the "stat" request comes in the middle of the night or on Sunday, this system is always in a "ready" condition; there is no startup/warmup problem.

D. Space Savings

The crowded, even cramped, conditions prevailing in most clinical laboratories need hardly be emphasized here. A laboratory which performs RIA either manually or with semiautomated equipment faces the separate problems of where to put the work stations for pipetting and reagent additions, the centrifuge, the water bath or heating block for incubation, the tube decanting or aspiration/wash station, and the gamma counter. Frequently, there also has to be a desk where data reduction can be done. All of these facilities not only take up valuable space; they also must often be placed in different places in the laboratory, around other pieces of equipment and work stations. They must be placed where the necessary services, such as 220 v electrical power or water supply are available. All of this results in a great deal of cross-traffic of personnel going from station to station; confusion, lost time, and accidental spills of samples and tracer can result. The problems, of course, become even more pronounced as the workload increases, and the pace becomes more hectic.

This answer to this space/function separation problem has been provided in the RIA laboratory by the development of an integrated reagent/instrumentation system in which all functions concerned in running an assay are carried out in one place, on one compact instrument.

E. Thyroid Profiling and Screening

The tests available on the CONCEPT 4® system permit the performance of an extensive thyroid profile. Typically, of course, a routine T_4 and T_3 Uptake will be run on a patient, often with the calculation of a Free Thyroid Index from the two results. The T_3 RIA and hTSH assays are most frequently used as followup tests when an abnormality is detected. In particular, the T_3 RIA test is often run when hyperthyroidism or T_3 toxicosis is suspected. On the other hand, hTSH levels are often evaluated in hypothyroid cases.

Of great significance is the applicability of the CONCEPT 4® system to neonatal hypothyroid screening programs. In the last few years it has been recognized that the mental retardation and other manifestations of cretinism are preventable if neonatal hypothyroidism is diagnosed early, and therapy undertaken during the first 3 months of life. For this reason a number of states and Canadian provinces have begun large-scale mass screening of neonates for the detection of the one hypothyroid case per 3,000 to 7,000 births.[4]

Clearly, the practicality of such mass screening programs depends heavily upon the availability of technology capable of handling the substantial workloads involved (estimated at 50,000 to 100,000 samples per year in the central laboratories to which specimens are sent in most jurisdictions). From the throughput analysis presented in Section IV., B. it can be seen that a highly automated system such as CONCEPT 4® has the necessary capability.

Typically, the screening test is a T_4 assay run on a dried filter paper spot (as discussed in Section III.,B.). The spots can be sent through the mail.

Those results which fall below a certain cutoff point (as discussed in Section II.C.1.c.) trigger a more extensive investigation of a given infant. This will often entail the performance of an hTSH assay on a serum sample or on the same filter paper spot from which the sample for the T_4 test was taken.

V. SUMMARY

In this chapter has been described a totally automated radioassay system which (1) because of high throughput enables laboratories to handle rapidly increasing workloads without increasing personnel or workspace; (2) frees available highly trained personnel from repetitive tasks to work on more challenging procedures; and (3) enhances performance by automating tedious, repetitive procedures which, when done manually, are an important source of laboratory error.

REFERENCES

1. **Dussalt, J. H.,** personal communication, 1978.
2. **Kolhouse, J. F., Kondo, H., Allen, N. C., Podell, E., and Allen, R. H.,** Cobalamin analogues are present in human plasma and can mask cobalamin deficiency because current radioisotope dilution assays are not specific for true cobalamin, *N. Engl. J. Med.*, 299, 785, 1978.
3. **Cooper, B. A. and Whitehead, V. M.,** Evidence that some patients with pernicious anemia are not recognized by radiodilution assay for cobalamin in serum, *N. Engl. J. Med.* 299, 816, 1978.
4. **Illig, R. and deVera Roda, R.,** Radioimmunological measurement of TSH in dried blood spots: a possible screening method for the detection of hypothyroidism in newborns, *Schweiz. Med. Wochenschr.*, 106, 1676, 1976.
5. Micromedic® Systems clinical validation data files.
6. **Rodbard, D.,** Statistical quality control and routine data processing for radioimmunoassays and immunoradiometric assays, *Clin. Chem.*, 20, 1255, 1974.

Chapter 3

AUTOMATED RADIOIMMUNOASSAY ON ARIA II®

L. R. Johnson

TABLE OF CONTENTS

I. INTRODUCTION

Radioimmunoassay, since its discovery in 1956 by Berson and Yalow,[1] has significantly improved the sensitivity and ease of performing quantitative and specific analysis on important substances of biological interest. A significant number of manual methods have appeared with variations in tracer and antibody properties that lead to important questions on the ability to compare laboratory results from one lab to another, or within a lab, since procedures, reagents, and technical skills are varied. Questions arise as to efficacy of one procedure over another, variations in standardization methods and data reduction. Separation of the free and bound fractions in RIA leads to shifts in binding depending on the separation method. Corrections for loss of analyte on extracted samples present extra work and more sources of error.

Automation offers improvements in removing a variable amount of technical involvement. Most automated systems that are available or are soon to be made available automate some of the existing manual procedures. Becton Dickinson Immunodiagnostics has offered a new approach in automation that conserves valuable antibody in a recyclable[2] process using ARIA II® (Figure 1). Antibody is covalently coupled to glass beads[3] or cellulose[4] and placed in an antibody chamber with an inner volume of about 50 μl. In a typical assay the sample is brought to the antibody chamber and allowed to bind as it passes through. The free antigen is counted and subsequently the bound antigen is eluted from the antibody and counted. A cycle takes 2 to 3 min, and several hundred cycles may be performed on a single antibody chamber. Unknowns and standards are run on the same antibody population which aids quality control. Aria II® has the capability of counting tritium as well as gamma emitting isotopes.

II. VERSATILITY OF ARIA II®

Not all assays are of the same type and the ability to be flexible offers the user additional procedures that may not be readily apparent. There are four types currently available on ARIA II®.

A. Simultaneous Injection of Tracer with Sample

Samples which contain tracer antigen with the nonradioactive sample antigen can readily be assayed by a simultaneous introduction of the tritium tracer and sample antigen to the antibody chamber as depicted in Equation 1.

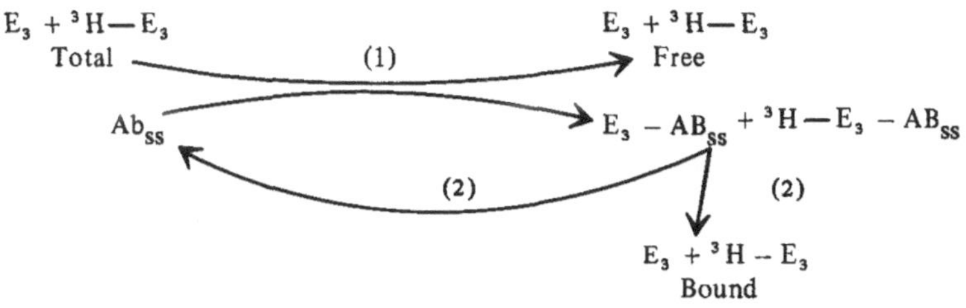

(1)

Reaction 1 illustrates a competitive binding of tracer antigen, in this case tritiated estriol, to estriol antibody covalently attached to a solid support. The sample 0.66 mℓ is pumped through the chamber at a rate of 1.1 mℓ/min. The free estriol is mixed with scintillation cocktail, 9.9 mℓ/min and counted. While the free is counted in a flow cell in the detector, the bound estriol is eluted from the antibody, reaction 2, with a 40% methanol solution in aqueous buffer at the same rate as the binding reaction.

The bound fraction is continuously mixed with liquid scintillation cocktail and counted. The antibody chamber and tubing is rinsed with adsorption buffer prior to initiating another cycle.

A typical elution curve, Figure 2, of the radioactivity passing through the detector illustrates good separation of the free and bound peaks. The counts are integrated for determination of the percent binding and the percent recovery. By placing an amount of tracer antigen in the sample prior to extraction, and comparing the radioactivity recovered with that obtained with controls that are not extracted, the percent recovery is determined and utilized for correcting the assayed value for losses during the extraction.

B. Competitive Protein Binding Assays

By automatically diluting the sample with tracer, allowing equilibration to take place and then separating the free and bound fractions with a specific antibody chamber, a unique competitive protein binding assay may be run. An example of this assay is the T_3 Uptake where serum, containing T_3 binding proteins such as thyroxine-binding globulin (TBG), are mixed with ^{125}I-triiodothyronine (^{125}I-T_3) according to reaction 1, Equation 2.

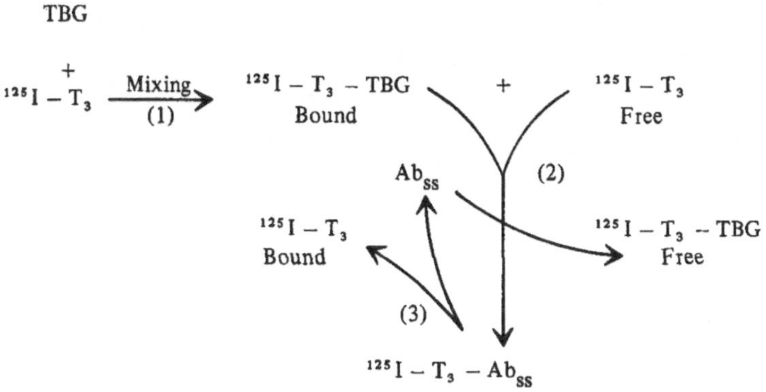

(2)

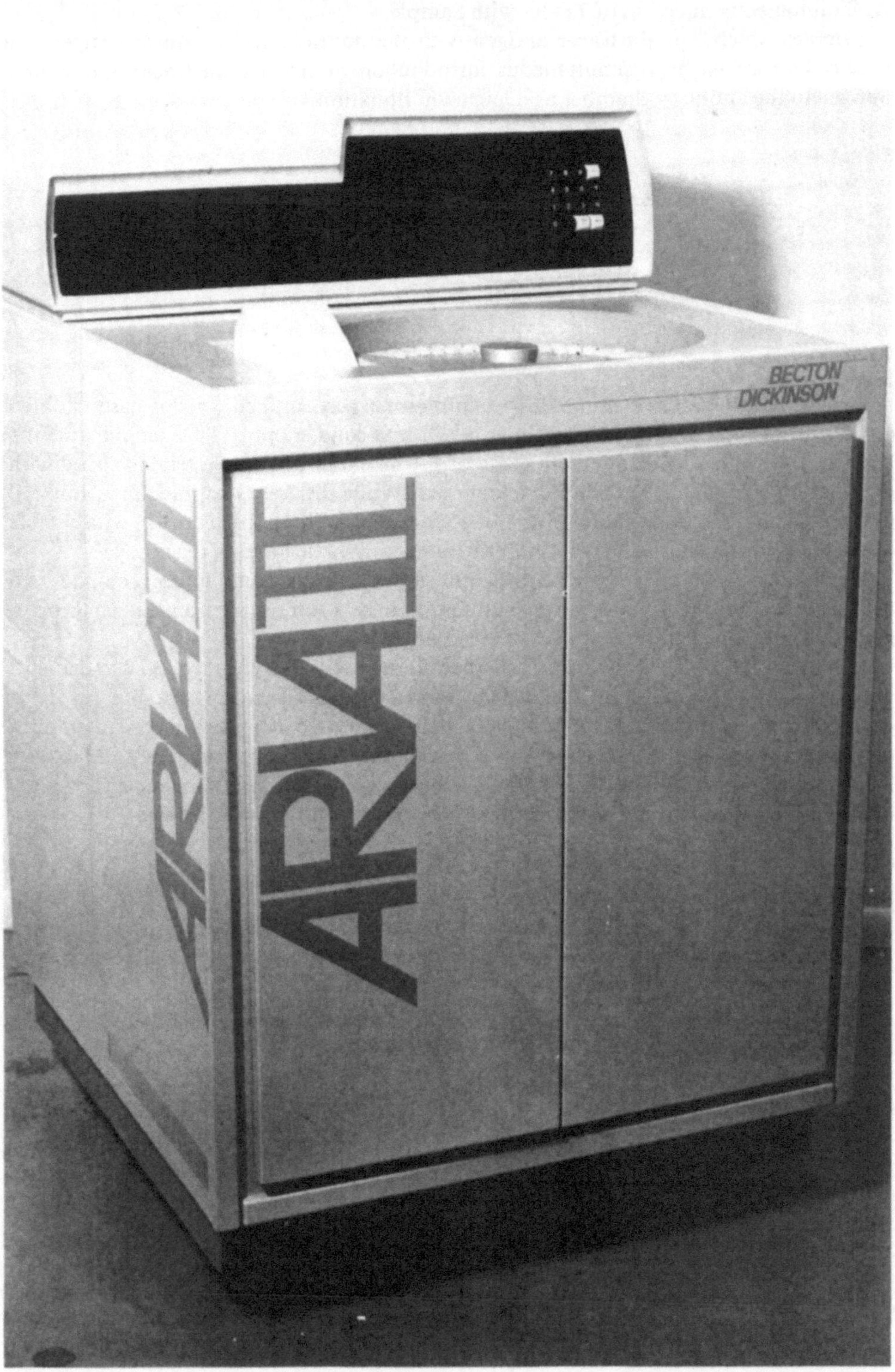

FIGURE 1. ARIA II®.

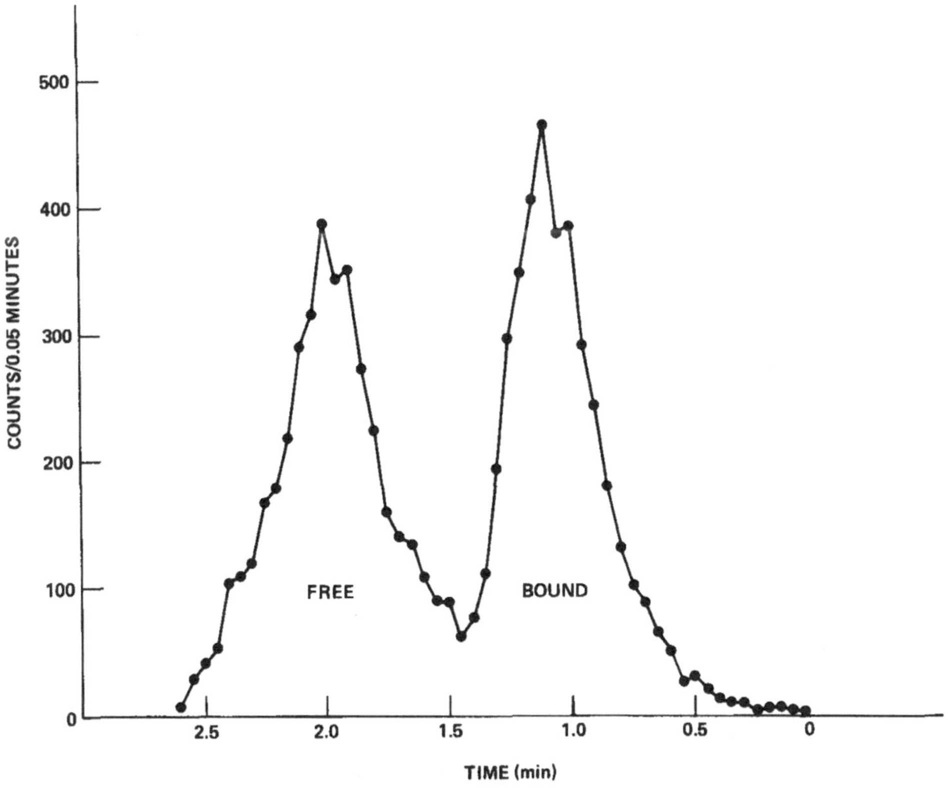

FIGURE 2. Estriol elution curve. A zero ng/mℓ tracer estriol solution was assayed. The time on the abcissa indicates time remaining in the cycle.

The equilibration of T_3 with TBG takes place rapidly. The free T_3 is taken up by the T_3 antibody on the solid support, reaction 2, while the T_3 bound to TBG is counted as the "free" fraction. The T_3 bound to antibody is eluted with a methanolic aqueous buffer, reaction 3, and counted as the bound fraction, the whole cycle taking 2.1 min.

C. Sequential Injection of Tracer

The serum is automatically diluted with diluent buffer about 30-fold as shown by the example for the thyroxine (T_4) assay. The diluent buffer contains releasing agents, such as ANS, to release T_4 from serum binding proteins. The diluted sample is then passed through the antibody chamber according to reaction 1, Equation 3.

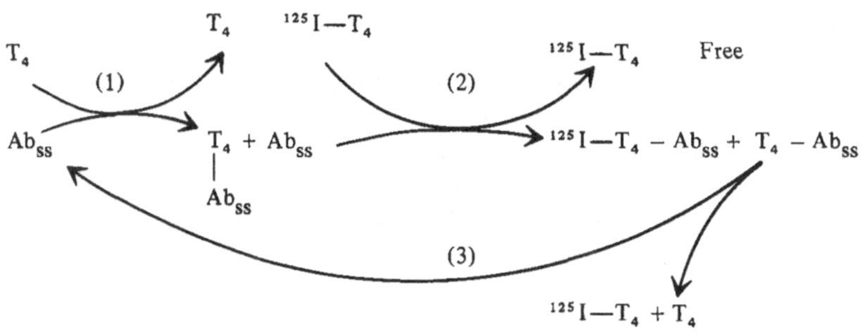

(3)

The free T_4 goes on to waste. Unreacted binding sites on T_4 antibody are then available for binding the tracer T_4 as shown in reaction 2. The free tracer T_4 is counted and the bound T_4 is eluted with a methanolic aqueous buffer and the radioactivity counted with the regeneration of antibody for another cycle, reaction 3. The sensitivity of the assay for T_4 is improved and the residual binding of tracer T_4 to the TBG in serum is eliminated. Thus the T_4 from the serum available for binding to the antibody is limited by the residual binding of T_4 to TBG not displaced by ANS in the competitive system established.

The sequential introduction of tracer is useful in eliminating serum related quenching of liquid scintillation counting of tritium, Table 1. A direct serum cortisol method was shown to be insensitive to quenching. Cortisol was measured in icteric serum samples on ARIA II®. The samples were diluted 30-fold automatically as described for T_4 and about 440 $\mu\ell$ of diluted sample were introduced to the antibody chamber. A second 50 $\mu\ell$ aliquot of the serum samples were diluted to 2.0 mℓ in the same sample buffer and liquid scintillation cocktail and counted in a Packard Liquid Scintillation Spectrometer.

Patient sample #5 had 12% quenching in the Packard with no apparent change in total counts on ARIA II®. All total counts on ARIA II® were within 99.3 and 102.8% of the mean count rate. In addition, an automatic channels ratio method for correcting the quench effect did not change the results. Thus the sequential injection of serum followed by the tritium tagged cortisol tracer effectively eliminates the serum quenching.

D. Double First Antibody (DFA)

In order to increase the sensitivity of a digoxin assay, a unique approach using a digoxin specific antibody in solution and digoxin specific antibody covalently bound to a solid support was devised.[5] ARIA II® aspirates 50 $\mu\ell$ of serum and transfers it to a second empty sample cup followed by 50 $\mu\ell$ of soluble antibody in diluent buffer and 100 $\mu\ell$ of digoxin tracer and 340 $\mu\ell$ of diluent buffer. The solution is mixed and allowed to incubate 18 min, reaction 1, Equation 4. The incubation solution is aspirated and passed through the antibody chamber.

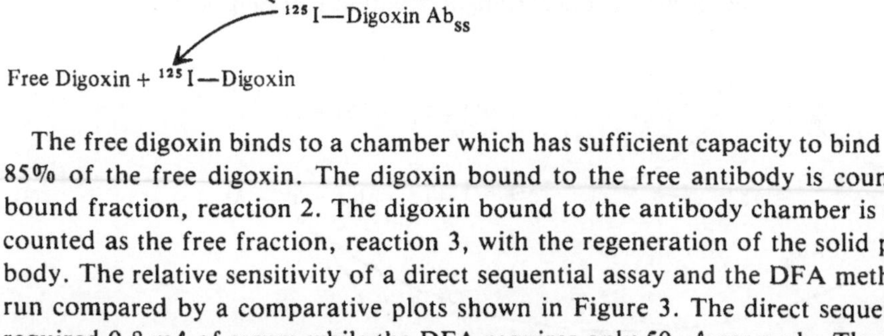

(4)

The free digoxin binds to a chamber which has sufficient capacity to bind more than 85% of the free digoxin. The digoxin bound to the free antibody is counted as the bound fraction, reaction 2. The digoxin bound to the antibody chamber is eluted and counted as the free fraction, reaction 3, with the regeneration of the solid phase antibody. The relative sensitivity of a direct sequential assay and the DFA method can be run compared by a comparative plots shown in Figure 3. The direct sequential assay required 0.8 mℓ of serum while the DFA requires only 50 $\mu\ell$ per cycle. The reproducibility was improved as the slope increased 40%.

Table 1
EFFECT OF BILIRUBIN QUENCHING ON SERUM CORTISOL WITH TRITUIM AS TRACER

			Cortisol ug/dℓ[b]			
			Uncorrected		Corrected[d]	
Patient #	Bilirubin mg/dℓ	Packard efficiency[a] %	$\bar{X}$	AV%DM[c]	$\bar{X}$	AV%DM[c]
1	3.2	99	28.2	6.0	28.8	4.8
2	3.2	98	31.4	3.5	33.1	3.6
3	10.8	91	40.6	0.2	43.1	3.7
4	6.9	95	38.5	3.9	36.5	1.1
5	15.2	88	31.2	1.4	31.8	2.3
6	5.5	95	18.3	6.5	17.1	4.6

[a] Packard quench on 50 μℓ of plasma in 2.0 mℓ of cortisol sample buffer containing 50,000 counts, in 18 mℓ of scintillation cocktail.
[b] ARIA II® Counting Efficiency = 99.3—102.8% for the paired samples with a total count of 9000 counts.
[c] AV%DM = Average percent deviation from the mean.
[d] Automatically corrected for quench by channels ratio method.

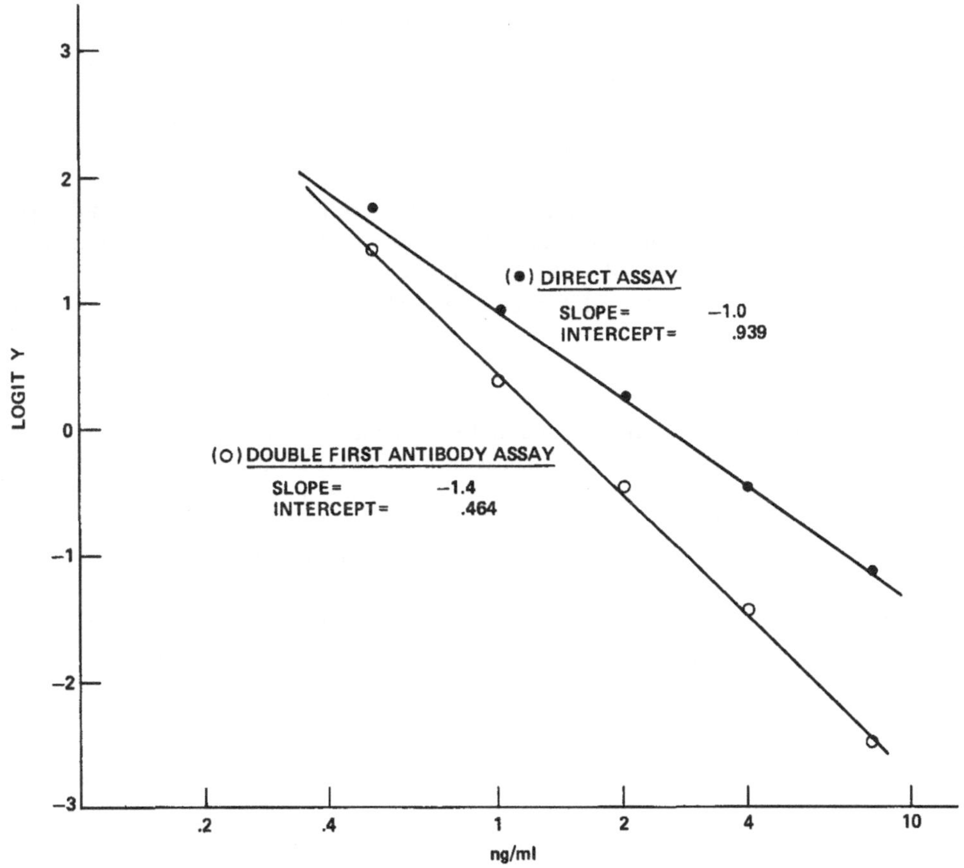

FIGURE 3. Digoxin standard curves. A direct sequential type assay was run with 1.0 mℓ of serum (closed circles) and a 2.9 min cycle time. A double first antibody assay with an 18 min incubation time and a cycle time of 2.0 min was run with 50 μℓ of serum (open circles).

III. DESCRIPTION OF ARIA II® AND ITS FUNCTIONS

A. Outline

ARIA II® is a self-contained instrument which consists of the following major modules:

1. Carousel for 120 sample cups
2. Aspirator
3. Pump and valve plumbing assembly
4. Beta and gamma scintillation detectors
5. Pulse height analyzer (PHA)
6. Control unit interface containing a mini floppy disc and a keyboard
7. Self-contained computer
8. Printer and CRT display

B. Description of a Typical Cycle

A description of a typical assay, e.g., thyroxine, and how the instrument performs the assay follows. A floppy disc containing specific instructions for control of the instrument for any particular assay available is inserted into and read by the floppy disc reader and the information is transferred to the computer. Information controlling pump rates and directions, valve positions, choice of detector, times for accumulation of free and bound counts, standard curve information and other control functions are transferred to the computer. Typically, there are six phases the instrument passes through for any run.

The first phase is designed to prime the pumps with appropriate solutions during which time the pumps are emptied and filled three times in about 7 min. This phase is not run if the same type of assay is run again.

The second phase positions the pumps for the initiation of the first cycle and aspirates the first sample. The antibody chamber is rinsed with elution buffer and adsorption buffer prior to being exposed to the first sample.

The third phase is utilized for DFA-type assays where incubations are set up for the first group of samples. The sample is aspirated and transferred to an empty sample cup followed by dispensing of free antibody and tracer and the resulting solution is mixed by aspiration and redispensing into the same incubation cup. By way of example, if the cycle time is 2 min and the incubation time is 18 min, then nine samples are set up in this phase of the run before proceeding to the next phase where incubations are set up and separation of bound and free take place simultaneously. The incubation phase is not used for the thyroxine assay being described. The fourth phase runs all but the first and last cycles. (A detailed description of this phase follows the sixth phase.)

The fifth phase is utilized for DFA assays only to gather the samples incubating after all the samples have been aspirated and incubations set up. Separation of the free and bound fractions takes place on the last samples.

The sixth phase performs the final assay for all types of assays. Generally, the last sample is aspirated during the next to the last cycle and this phase is used to introduce the sample to the antibody chamber and complete the analysis on the last sample. Upon completion of this cycle the instrument goes to an idle chamber in mode where adsorption buffer is passed through the chamber at a slow rate.

The details of phase four of the thyroxine assay illustrates the movement of buffers, valves, and critical functions of the assay. Figure 4 illustrates the plumbing of ARIA II®. Table 2 provides the actions that occur during this phase. This phase cycles re-

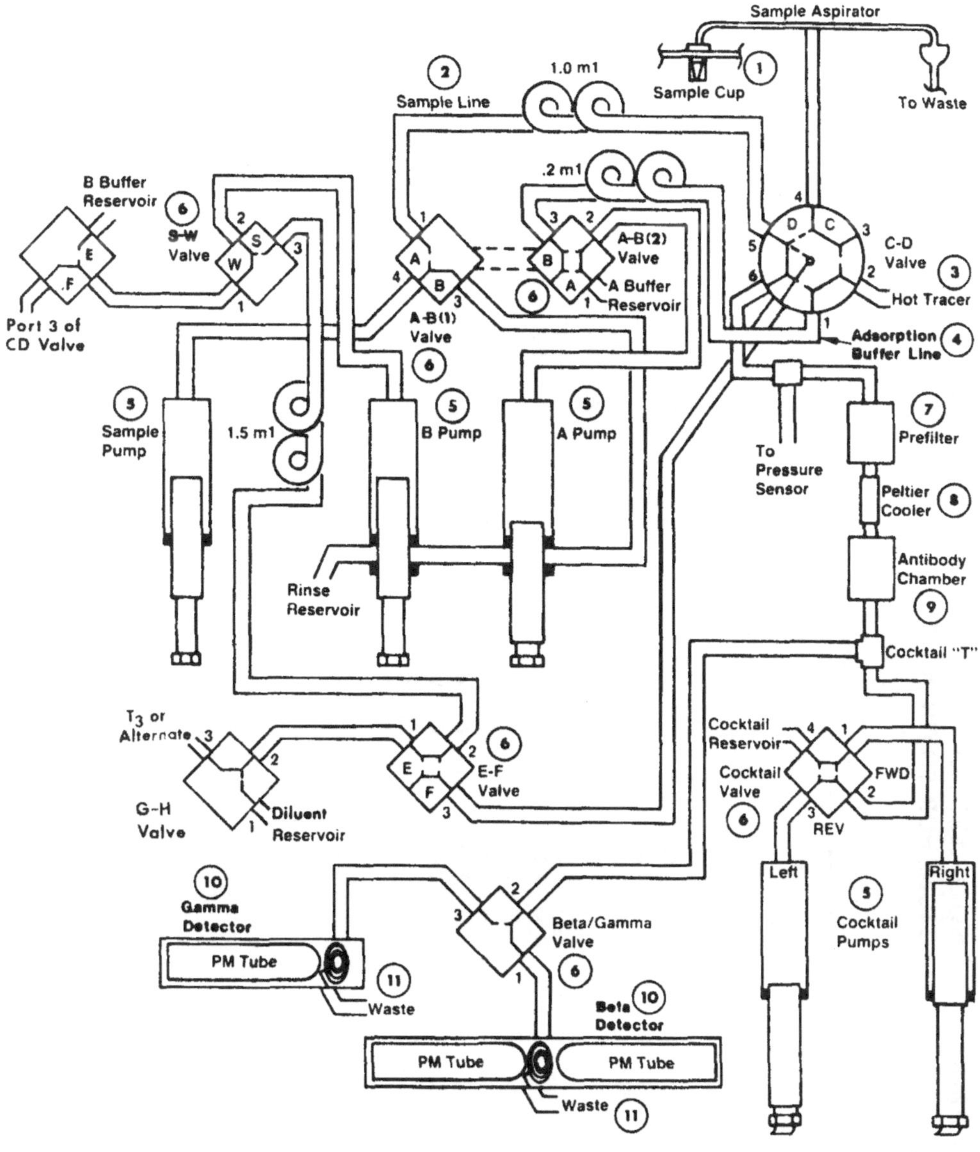

FIGURE 4. ARIA II® pump and valve schematic.

petitively during a run. As period A approaches, the sample has already been introduced to the chamber, and the free thyroxine is about to appear at the gamma detector to be counted. Adsorption buffer is pumped by the A pump at a rate of 3.30 mℓ/min through the AB valve to port 1 of the CD valve; it leaves the CD valve at port 6 and passes through the prefilter, Peltier cooler, and antibody chamber. The left cocktail pump dispenses detergent solution which mixes with the adsorption buffer to prevent T_4 from binding to the flow-cell tubing in the detector. Free counts are accumulated. Diluent buffer is pumped by the B pump through the SW, EF, and CD valves, to the aspirator tubing in the rinse position, to waste. This positions the elution buffer in the tubing between port 3 of the EF valve and port 5 of the CD valve. Rinse solution is also pumped to the aspirator tubing with the S pump through the AB and CD valves to waste.

Table 2
INTERNAL PROGRAMMING OF ARIA II® PUMP, VALVE AND OTHER FUNCTIONS FOR THYROXINE

	Pump direction[a]				Pump rate				Valve position						
Period	A	B	S	C	A	B	S	C	A-B	C-D	S-W	E-F	Aspirator[b]	Time min.	Count
A	o	o	o	i	3.30	3.30	6.10	1.00	A	D	S	E	R	0.10	Free
B	o	o	o	i	3.30	0.00	6.10	1.00	A	D	S	E	R	0.23	Free
C	o	o	o	o	3.30	0.00	6.10	1.00	A	D	S	E	R	0.23	Free
D	i	o	i	o	9.99	3.30	6.80	1.00	B	C	S	E	R	0.25	Free
E	i	o	i	o	9.99	3.30	6.80	1.00	B	C	S	E	R	0.25	Bound
F	o	i	i	i	3.30	8.87	0.00	1.00	A	D	S	F	A	0.15	Bound
G	o	o	i	i	3.30	1.20	3.01	1.00	A	D	S	E	A	0.10	Bound
H	o	o	i	i	3.30	5.10	5.27	1.00	A	D	S	E	A	0.15	Bound
I	o	i	o	i	3.30	9.90	0.60	1.00	A	D	W	E	A	0.05	Bound
J	i	o	o	o	9.99	3.30	0.00	1.00	B	C	S	E	R	0.05	Bound
K	i	i	o	o	0.45	9.99	1.66	2.64	A	C	W	E	R	0.20	Bound
L	i	i	o	i	0.45	9.99	1.66	2.64	A	C	W	E	R	0.20	Bound
M	o	o	o	i	1.66	0.00	6.10	2.64	A	D	W	E	R	0.15	
N	o	o	o	o	1.66	0.00	6.10	2.64	A	D	W	E	R	0.15	
													Cycle Time	2.26	

[a] O = out, i = in.
[b] R = Rinse Position, A = Aspirate Position.

During period B the adsorption buffer continues to be pumped, and the elution buffer in the aspiration tubing is rinsed with 1.3 mℓ of solution by the S pump. The cocktail pump (C) is reversed during period C with the left pump filling and the right pump discharging gamma cocktail. The adsorption buffer continues to rinse the antibody chamber of free counts.

Elution of the bound thyroxine from the antibody chamber is begun in period D. Elution buffer is pumped through the SW, EF, and CD valves. At the CD valve, the elution buffer enters the central port which is internally connected to port 5 and emerges at port 6 with the CD valve in the C position. The free counts continue to accumulate as the bound counts approach the detector at the end of this period. The A and S pumps are filled with adsorption buffer and rinse solution, respectively.

The eluted bound thyroxine appears at the detector during period E and accumulates through period L. The B pump aspirates 1.33 mℓ of sample buffer from the diluent reservoir into the 1.5 mℓ holding tubing in period F, while the antibody chamber is rinsed with adsorption buffer through periods F-I.

In period G 181 $\mu\ell$ of sample are aspirated from the sample cup as 120 $\mu\ell$ of sample buffer are moved from the EF valve to port 5 of the CD valve. The sample is diluted 32-fold as the S pump aspirates sample and diluent buffer at the rate of 5.27 mℓ/min with the B pump providing the diluent buffer at the rate of 5.10 mℓ/min to port 5 of the CD valve.

During period I, the sample pump is reversed and 30 $\mu\ell$ of diluted sample is pumped into the aspirator tubing to properly seat the pump. The B pump is filled with elution buffer.

Additional diluent buffer is pumped up to the CD valve with the B pump during period J and into the prefilter line to the antibody chamber prior to introducing the sample.

The diluted sample, 660 $\mu\ell$ equivalent to 20 $\mu\ell$ of serum in buffer containing ANS to release T_4 from TBG, is introduced to the antibody chamber in periods K and L. At the same time 180 $\mu\ell$ of ^{125}I-T_4 is aspirated by the A pump through port 2 of the CD valve to the holding tubing between the CD valve and the AB valve. Two periods are required so the cocktail pump direction can be reversed since it is a reciprocating pump. At the conclusion of period L, the bound counts are accumulated and the results of the assay are printed.

The tracer is introduced in periods M and N at the rate of 1.66 mℓ/min for 0.3 min. At this point a new sample is in process and the phase is reinitiated. The overall cycle time being 2.26 min.

C. Detector and Pulse Height Analyzer

1. Gamma Detector

The ^{125}I gamma rays emitted by the tracer are detected by a standard photomultiplier tube coupled to sodium iodide thallium activated crystal encased in aluminum. The counts are transferred to 256 channel pulse height analyzer and accumulated for each free and bound portion of the cycle as described above. The counts for several cycles are accumulated and stored in the computer for PHA spectrum recall at the end of a run for quality control purposes. A typical pulse height analyzer spectrum for I-125 is illustrated in Figure 5. The absolute efficiency of the detector in this example was 68% as calculated by the Eldridge method.[6]

There is ample room available to the right of the sum peak to count Co-57 simultaneously with I-125 in dual isotope assays.

2. Beta Detector

The beta liquid scintillation detector consists of two photomultiplier tubes mounted

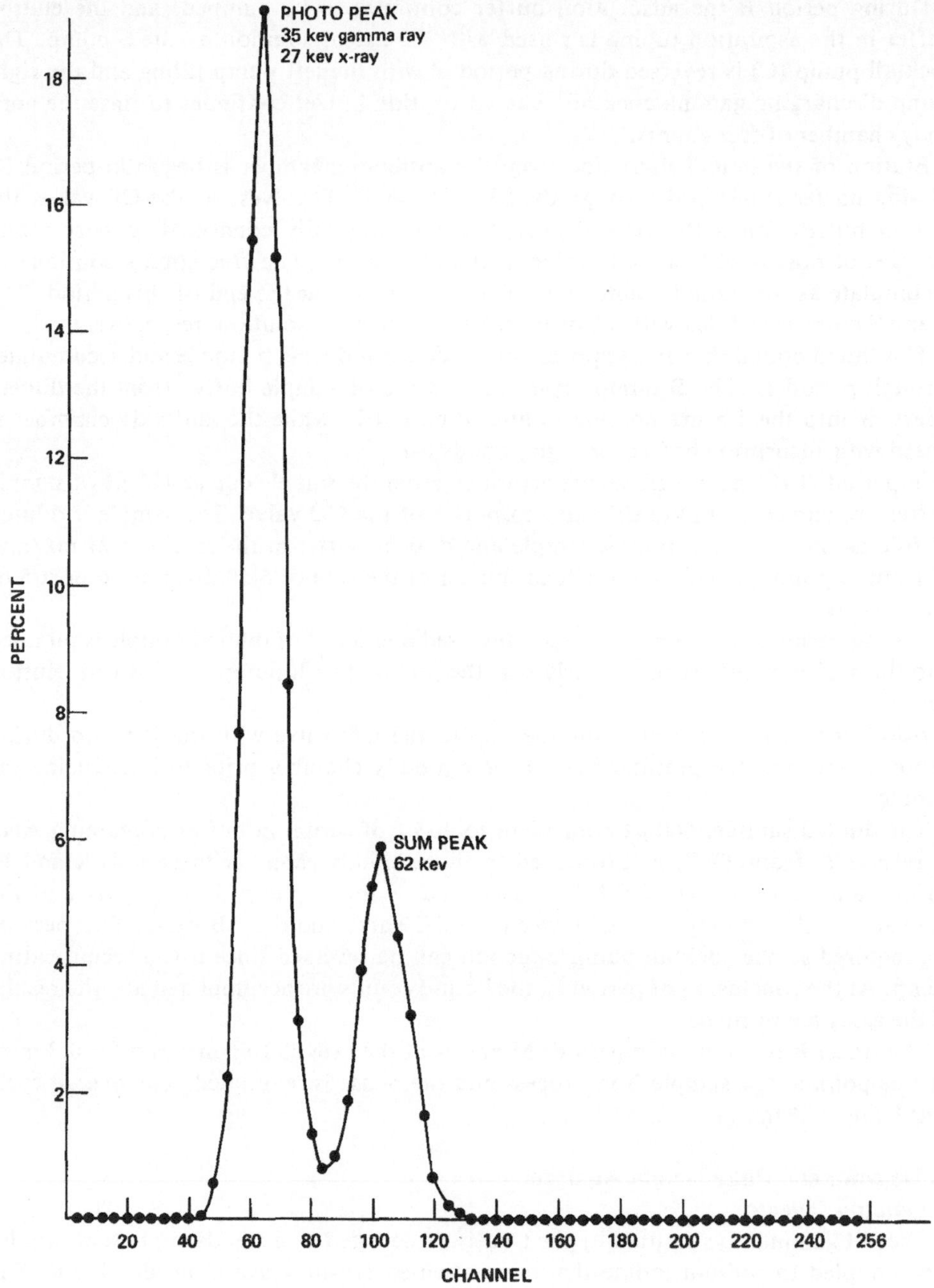

FIGURE 5. Iodine −125 PHA spectrum on ARIA II®. The percent of total counts was plotted for the 256 channels available. Four channels were added for each data point.

opposed to each other with a flow cell in between. The flow cell consists of a spiral coil of teflon tubing of approximately 3.5 mℓ volume. Aqueous samples are diluted ninefold continuously with liquid scintillation cocktail which solubilizes the two solutions.

Tritium is detected with an efficiency of 30% using coincidence techniques. The computer accumulates the counts due to the tritium decay for each assay, and as in the case for gamma emitting isotopes, accumulates counts over several cycles so that a pulse height analysis spectrum (PHA) may be printed at the end of a run. A typical PHA spectrum for tritium is illustrated in Figure 6.

D. Peltier Cooler

Temperature control of the solutions entering the antibody chamber are controlled to 0.1°C by a thermoelectric device. The flow rates and characteristics of the heat exchange are sufficient to control the temperature anywhere between 20 and 30°C as selected and stored in the floppy disc for each assay. Since the instrument generates some heat, the temperature can be controlled with an ambient temperature range of 18 to 30°C.

E. Elution Curves

At any time during a cycle, the print mode may be changed to print an elution curve without affecting the performance of the assay. A typical elution curve for thyroxine is illustrated in Figure 7. Free counts are accumulated from 1.95 to 1.20 min and bound counts from 1.20 min to the end of the cycle. The low counts at the first of the elution curve occur at the end of the cycle where neither free nor bound counts are accumulated.

F. Data Reduction

The percent binding data for the standards are used to determine the constants in an equivalent Hill equation by means of a "matched response" curve fitting algorithm.[7,8] The Hill equation as applied to RIA is equation 5 as follows:

$$\frac{B}{Bo} = \frac{CK^m}{1 + CK^m} \qquad (5)$$

Where B = Percent of total bound radioactivity, Bo = B for the zero standard with tracer only, C = A constant (where ln C is the intercept of the logit-log curve), m = A constant (slope of logit-log curve), K = Concentration of antigen.

Equation 5 may be rearranged to Equation 6.

$$CK^m = \frac{B/Bo}{1 - B/Bo} = \frac{B}{Bo - B} \qquad (6)$$

Nonspecific binding (Bm) may be determined from the standards themselves, which establishes an effective range for the standard curve, Equation 7.

$$\frac{B - Bm}{Bo - B} = CK^m \qquad (7)$$

The familiar logit-log form of the Equation 8 can be derived by taking the logarithms of Equation 7.

$$\ln \frac{B - Bm}{Bo - B} = m \ln K + \ln C \qquad (8)$$

The slope and intercept of the logit-log Equation 8 can be solved readily if the Bo and Bm are accurately known. This can be done by making several replicate (10 or more) determinations of Bo and Bm, or an alternate method would be to assume the

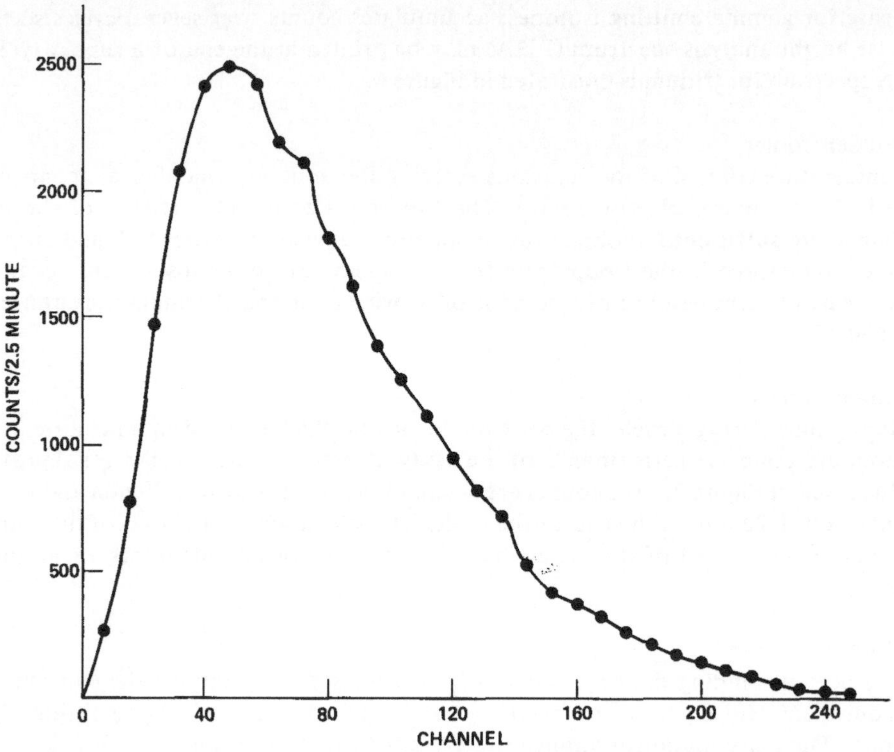

FIGURE 6. Tritium PHA Spectrum® on ARIA II®. Tritium counts were accumulated for 2.5 min and then the counts were printed for 256 channels. Four channels were summed for each point.

standards may determine Bo and Bm uniquely. Thus there are four unknowns to be determined from five serum standards on ARIA II®.

By rearranging Equation 7 into the hyperbolic form, Equation 9, these constants may be determined by

$$K^{-m} (B - Bm) - C(Bo - B) = 0 \quad (9)$$

Using a least squares approach, as provided in Equation 10,

$$K^{-m} (B - Bm) - C(Bo - B) = \delta i \quad (10)$$

where the $\Sigma^{n}_{i=1} \delta i^2$ is minimized on several samples. By selecting a reasonable value for m; Bo, Bm and C may be solved by taking partial derivatives of Equation 10 with respect to the three constants. The derivation of the complex equations is available.[7] Once the Bo, Bm, and C are determined, the slope is adjusted, and the process iterated to minimize the difference between the actual and calculated values for the standards.

By experience, outliers in a standard curve are usually disregarded by an experimenter, otherwise the curve would be obviously skewed. In the standard least squares approach as described above, outliers exert a disproportionate effect on the curve. By placing less weight on the standard curve points that are further from the expected distribution, their effect on the standard curve parameters are minimized. This is achieved using the "matched response" technique with an appropriate weighting function.[8]

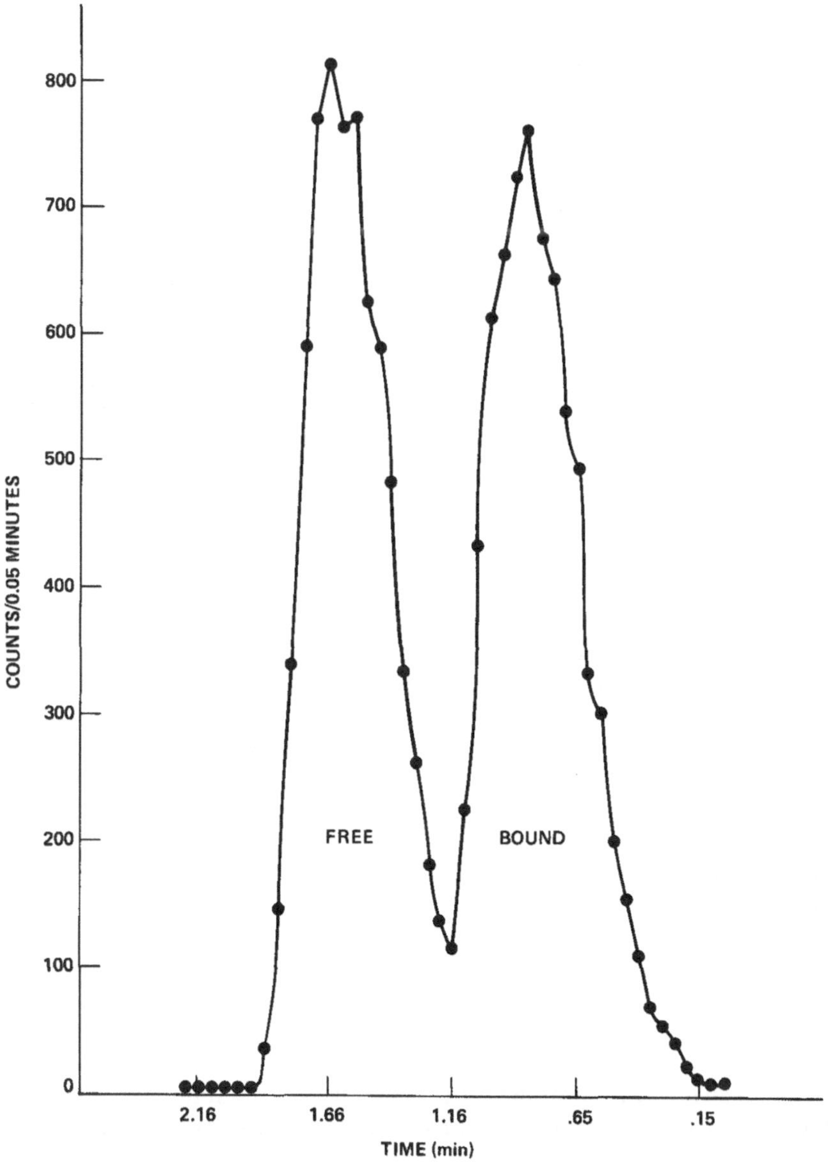

FIGURE 7. Thyroxine elution curve. The accumulated counts for 0.05 min are plotted against remaining time in the cycle.

G. Self Diagnostic Features

1. Safety Messages

The instrument monitors several functions for protection. A vapor sensor monitors the air within the reagent and waste storage area for solvent vapors at less than 0.1 of the lower explosive level. This may occur if a full reservoir sensor in the waste reservoir malfunctioned coincident with being overfilled or a leak developed in critical tubing or fittings. In the event solvent vapors are detected the instrument immediately shuts down in a standby mode.

2. Malfunctions

Reagent volume consumption is monitored for each run and the operator is informed of insufficient reagents when the condition occurs at the start of a run. If the system is by-passed by the operator, the instrument will go to standby if the reservoirs are empty or the waste receptacle is full. Excessive back pressure in the tubing prior to the antibody chamber is sensed and shuts down the instrument.

3. Assay Functions

Alert messages are activated when the counts are less than the minimum acceptable for the assay or when the CV of the mean of total counts is greater than 2.0. Unknowns that calculate outside the range of the standard curve are so indicated with $<$ or $>$ symbols.

4. Recalculation of Data

If the printer manfunctions during a cycle, or runs out of paper an alert will sound and appear. The data are stored in the core memory and are retrievable even if the power fails in the meantime. The run may be recalculated and appear on the CRT and printed.

IV. ASSAY PERFORMANCE

A. Available Assays on ARIA II® — February 1979

Progesterone and estriol must be extracted prior to analysis. The other assays utilize serum or plasma directly. The volume of sample placed in the sample cup is not critical but must be sufficient for handling (about 1 mℓ). The actual amount of serum introduced into the antibody chamber is considerably less. Generally six or seven standards are run to calibrate the instrument followed by two controls and then unknowns. A summary of the test parameters appears in Table 3.

B. Extraction Assays

1. Estriol

a. Extraction Procedure

Twenty $\mu\ell$ of estriol (2,4,6,7-^{3}H) containing 180 nCi are added to 0.50 mℓ of serum. This solution is extracted with 10 mℓ of 3:2 ethylacetate-hexane solution. Five mℓ of the extract are evaporated to dryness and the dried residue is dissolved in 1.0 mℓ of sample buffer at 40°C for 15 min. The resulting solution is assayed. Standards containing an equivalent concentration of tracer are run to establish the total count for recovery calculations. A dilution factor of 4 is utilized in unknown calculations.

b. Standard Curve

A typical estriol standard curve is represented in Table 4. The percent binding and total counts are presented for each standard. Based on extensive data the best slope for this assay was found to be −1. The intercept, Bo and Bm were determined as described earlier. The confidence factor, CF, of 0.95 indicates that the mean deviation of the value of the calculated points on the standard curve from the actual points was 5%.

c. Accuracy

Accuracy was assessed by linear regression correlations with manual kits, recovery experiments, linearity on dilution and specificity studies. The antibody was obtained from rabbits utilizing estriol-6-carboxymethyl oxime conjugated to rabbit serum albumin.

Linear regression correlation — Twenty-nine 2nd or 3rd trimester pregnancy samples

Table 3
ASSAYS AVAILABLE ON ARIA II® AND TEST PARAMETERS

Assay	Sample size ($\mu\ell$)	Sample preparation	Incubation time (min)	Range	Sample injected ($\mu\ell$)	Cycle time (min)	Type of tracer
Cortisol	200	—	—	1—64 ng/dℓ	13	2.30	^{3}H
Progesterone	500	Organic extraction	—	1—32 ng/dℓ	138	2.40	^{3}H
Estriol	500	Organic extraction	—	1—32 ng/dℓ	165	2.66	^{3}H
Macro-Thyroxine (Auto-Dilution)	200	—	—	1—32 ng/dℓ	21	2.26	^{125}I
Micro-Thyroxine	20	Manual dilution	—	2—32 ng/mℓ	16	2.40	^{125}I
T_3 Uptake	200	—	—	30%U—Normal	12	2.32	^{125}I
Digoxin	150 (duplicate)	—	18	0.5—6 ng/dℓ	33	2.00	^{125}I
	75 (singlet)	—					

Table 4
ESTRIOL STANDARD CURVE

A. Actual Data

#	Concentration (ng/mℓ)	%Bound	Counts
1	0.00	46.6	8643
2	0.25	42.8	8940
3	0.50	39.1	9141
4	1.00	33.1	8990
5	2.00	26.3	9069
6	4.00	20.2	9028
7	8.00	13.8	9167

B. Standards Estimated on Self Curve

Concentration (ng/mℓ)		
Actual	Calculated	Difference (%)
0.00	0.01	
0.25	0.23	7.7
0.50	0.49	2.6
1.00	1.05	4.8
2.00	2.07	3.5
4.00	3.78	5.5
8.00	8.22	2.7

C. Curve Parameters

Mean Total Count	8997
CV	1.95
Slope	−1.0
Intercept	0.747
Bo	46.9%
Bm	5.3%
Confidence Factor	0.95

were analyzed on Aria II® and by two manual estriol (^3H) kits (I and II) and one manual (^{125}I) kit (III). The manual kit results were plotted against ARIA II® results and the slope characteristics were determined, Table 5.

Recovery and linearity — Estriol was added to estriol free serum, extracted and analyzed, Table 6. Another sample containing 32 ng/mℓ estriol was diluted proportionately and assayed on ARIA II®, Table 7.

Specificity — Several potentially interfering steroids were analyzed for cross reactivity at the 50% of Bo, prior to extraction, Table 8, Estriol-3-sulfate and estriol-3-glucuronide had 46 and 20% cross reactivity. By extracting serum samples fortified with these two estriol metabolites since they are present in high concentrations in pregnancy serum,[7-10] the cross reactivity was less than 1%. Thus the assay measures unconjugated estriol rather than "total" estriol which would be highly variable depending on the concentrations of the relative estriol conjugates. By proper hydrolysis techniques, total estriol may be determined in urine.

d. Reproducibility

Two pooled serum samples were analyzed for intra- and inter-assay reproducibility, Table 9.

Table 5
ESTRIOL CLINICAL CORRELATIONS

Reference method	Correlation coefficient	Slope	Y-Intercept
Kit I	0.978	0.819	0.13
Kit II	0.954	1.016	1.08
Kit III	0.969	0.993	1.08

Note: Reference Method results for 29 samples were plotted on the Y-axis against ARIA II® results on the X-axis.

Table 6
ESTRIOL RECOVERY DATA ON ARIA II®

Estriol added, ng/m*l*	4	8	16
No. of determinations	5.0	5.0	5.0
Estriol found, ng/m*l*	4.15	8.63	15.86
SD	0.37	0.80	1.36
% CV	8.9	9.2	8.5
% Recovery	103.0	107.0	99.0

Note: Estriol was added to estriol free serum.

Table 7
ESTRIOL LINEARITY ON ARIA II®

No. of determinations	% Dilution	Mean, ng/m*l*	SD	% C.V.
5	50	32.05	1.05	3.3
5	40	32.51	0.76	2.3
5	30	31.67	0.67	2.1
5	20	33.32	0.81	2.4
5	10	30.15	2.77	9.2
	Average =	31.94	1.21	3.9

Note: A sample containing 32 ng/m*l* diluted prior to assay.

2. Progesterone

a. Extraction Procedure

Twenty μ*l* of progesterone (1,2,6,7-^{3}H) containing 220 nCi are added to 0.5 m*l* of serum from nonpregnant samples (pregnancy samples are diluted 20-fold with water before sampling). The serum is extracted with 10 m*l* of Nanograde Hexane for 20 to 30 min on a gentle shaker or tube rocker. Five m*l* of extract are transferred to clean glass culture tubes and the solvent is removed by evaporation in a water bath at 45°C. The dried residue is dissolved in an aqueous buffer containing 0.1% bovine serum albumin in a phosphate-saline buffer at pH 7.4 while covered in a 45°C water bath

Table 8
ESTRIOL SPECIFICITY

Compound	%Cross-Reactivity
Estriol (3, 16α 17β-triol)	100.0
Estriol-3-sulfate	46.0
Estriol-3-(β-D-glucuronide)	20.0
Estrone	1.0
Estriol-17β-(β-D-glucuronide)	0.4
Estriol-16α-(β-D-glucuronide)	0.3
Estradiol	0.02
Androstenedione	0.001
Testosterone	0.01
Progesterone	0.01
Deoxycorticosterone	0.01
11-Deoxycortisol	0.01
Cortisol	0.01

Note: Aqueous solutions were analyzed on ARIA II®. The cross-reactivity was at 50% of the Bo binding. Extraction of the estriol-3-sulfate and estriol-3-(β-D-glucuronide) reduced the cross-reactivity to less than 1%.

Table 9
ESTRIOL REPRODUCIBILITY

Sample	Medium	High
Intra-Assay		
No. of determinations	10.0	10.0
Mean, ng/mℓ	5.60	12.95
SD	0.22	0.63
%CV	3.8	4.8
Inter-Assay		
No. of determinations	18.0	18.0
Mean, ng/mℓ	5.78	12.87
SD	0.45	1.05
%V	7.8	8.2

for 1 hr. Appropriate extraction controls are run to ensure extraction efficiency and purity of the hexane. Impure hexane may be purified by passing through a silica gel column.

b. Standard Curve

The parameters for the standard curve are given in Table 10, and they have the same meaning as for the estriol standardization. Reproducibility in sampling is shown by the reproducibility of the total counts. For 10,000 counts, the CV would be expected to be 1 for radioactive decay.

c. Antibody Chamber Stability

Standard curve samples were run on a single chamber for over 500 cycles to study the stability of the chamber, Figure 8. By calibrating each run the effects of small variations are minimized.

Table 10
PROGESTERONE STANDARD CURVE

A. Actual Data

#	Concentration (ng/mℓ)	%Bound	Counts
1	0.000	57.2	11,608
2	0.125	51.4	11,697
3	0.250	47.6	11,830
4	0.500	41.3	11,935
5	1.000	32.3	12,160
6	2.000	24.9	11,968
7	4.000	17.2	12,070
8	8.000	13.6	11,900

B. Standards Estimated on Self Curve

Concentration (ng/mℓ)		Difference
Actual	Calculated	(%)
0.00	0.008	*****
0.12	0.140	12.1
0.25	0.247	1.3
0.50	0.481	3.8
1.00	1.024	2.4
2.00	1.923	3.9
4.00	4.357	8.9
8.00	7.829	2.1

C. Curve Parameters

Mean total count	11,896
CV	1.44
Slope	−1.0
Intercept	0.004
Bo	57.3%
Bm	8.0%
Confidence factor	0.95

d. Progesterone Accuracy

Progesterone levels in female follicular serum samples are less than 1 ng/mℓ, between 1.0 and 30 for the luteal phase, and 30 and 300 for pregnancy. Male values are less than 0.8 ng/mℓ.

Clinical correlations with reference methods — Between 52 and 67 clinical samples were assayed by a commercial kit and by a reference laboratory. Excellent correlation was achieved between the methods, Table 11. The manual kit values agreed with ARIA II®, but the reference laboratory values were only 72% of the ARIA II® or manual kit values. Extraction losses were ignored by the reference lab method. This corresponds to extraction recoveries of 80 to 90% when samples are extracted by the ARIA II® method.

Recovery — Progesterone was added to a serum sample containing 0.58 ng/mℓ progesterone. Recoveries ranged from 89 to 96% when 2, 4 and 8 ng/mℓ were added prior to extraction, Table 12. Extraction losses as determined by recovery counts were used to correct the result for these losses automatically.

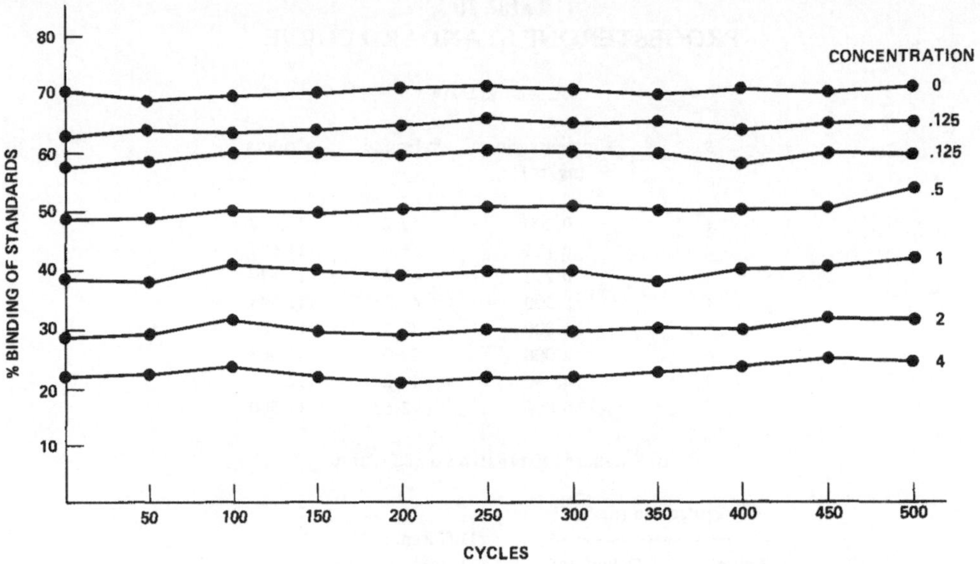

FIGURE 8. Progesterone antibody chamber stability. The percent binding for progesterone standards (ng/m*l*) were plotted with number of cycles as new standard curves were run over a period of 500 cycles.

Table 11
PROGESTERONE CLINICAL CORRELATION WITH REFERENCE METHODS

X-Axis	Y-Axis	Correlation coefficient	Slope	Y-Intercept	N
ARIA II®	Manual Kit	0.99	0.97	2.1	67
ARIA II®	Reference lab	0.98	0.73	2.1	57
Man. kit	Reference lab	0.98	0.72	1.7	52

Table 12
PROGESTERONE RECOVERY IN SERUM

Progesterone added (ng/m*l*)	Assayed value (ng/m*l*)	SD	%CV	N	Percent recovery
0	0.58	0.05	8.6	5	100
2	2.35	0.14	5.9	5	89
4	4.43	0.12	2.6	5	96
8	8.27	0.37	4.5	5	96

Average Recovery = 94%

Linearity on Dilution — Serum containing 3.2 ng/m*l* was diluted, extracted and assayed as indicated, Table 13. The linearity was excellent, with a mean of 3.3 and a CV of 7.8 for all of the dilutions.

Specificity — 5-Dihydroprogesterone and pregnenolone had significant cross reaction with aqueous standard, Table 14. However, upon extraction, the cross reaction decreased to less significant levels. Since the antibody was induced in rabbits with progesterone-6-carboxymethylene thioether coupled to bovine thyroglobulin the A and

Table 13
PROGESTERONE LINEARITY ON DILUTION

Percent serum	Calculated value (ng/ml)	SD	%CV	N
5	3.42	0.20	11.4	5
10	3.64	0.23	6.4	5
20	3.41	0.42	6.7	5
30	3.14	0.38	4.0	5
40	3.06	0.68	5.5	5
50	2.98	0.55	3.7	5
Average	3.28	0.25	7.8	6

Table 14
PROGESTERONE SPECIFICITY

	Percent cross reactivity at 50% of Bo	
Compound	Before extraction	After extraction
E_1	+n0.1	
E_2	+n0.1	
Androstenedione	0.1	
Corticosterone	0.3	0.1
5α-Dihydroprogesterone	89.0	10.0
20α-Dihydroprogesterone	3.0	0.4
Cortisol	0.1	
Testosterone	0.1	
Pregnenolone	44.0	6.0
17α-Hydroxyprogesterone	0.1	0.4
11-Deoxycortisol	3.0	
DOC	4.0	2.0
Cortisone	0.1	0.1

B rings of progesterone were expected to confer less specificity than the C and D rings or the side chain.

e. *Progesterone Precision*

Groups of ten replicates of low, medium, and pregnancy samples were analyzed for intra- and inter-assay reproducibility with CVs of less than 9 for all but the lowest sample in the inter-assay group, Table 15.

C. Direct Serum Assays on ARIA II®

1. Cortisol

a. *Dilution of Sample*

Cortisol in serum is predominantly bound to corticosteroid binding globulin (CBG), and must be dissociated prior to analysis. By diluting serum with a citrate-phosphate buffer at pH 3.5, cortisol does not bind to CBG but does bind to antibody. Approximately 200 μl of serum are placed in a sample cup. An aliquot of the sample is diluted similarly to T_4 described earlier and assayed.

Table 15
PROGESTERONE PRECISION

	Sample		
Parameter	Low	Medium	Pregnancy
	Intra-Assay		
Mean, ng/mℓ	0.58	12.3	124.0
SD	0.04	0.6	10.0
CV	7.8	4.9	8.1
N	10.0	10.0	10.0
	Inter-Assay[a]		
Mean, ng/mℓ	0.57	13.7	142.0
SD	0.16	0.74	8.8
CV	28.6	5.4	6.2
N	15.0	15.0	15.0

[a] Duplicates were run during each run.

b. Standard Curve

The reproducibility of the total counts confirm accuracy in dispensing the radioactive tracer, Table 16. A confidence factor of 0.95 indicates reasonable fit for the standards. A Bm of 11.6 indicates some nonspecific binding; however, this is probably related to the potential range of the standards if higher concentrations were used.

c. Accuracy

Clinical correlation — The correlation coefficients with two manual kits were at least 0.97. One kit had a 12% negative bias to the ARIA II® results, while the other kit agreed, Table 17.

Cortisol linearity — A high cortisol pool was diluted up to fivefold manually in aqueous buffer prior to placing on the instrument. The experimental values corrected for dilution agree, Table 18.

Specificity — Antisera to cortisol-3-carboxymethyloxime coupled to bovine serum albumin was produced in rabbits. The principle cross reacting steroids are similar to cortisol, Table 19. Prednisone and prednisolone interference may be eliminated by not collecting samples within 48 hr of drug administration. The relatively low cross reaction by 11-deoxycortisol allows metyrapone suppression tests to be run. High levels of 6-hydroxycortisol are found in neonates. Other steroids would not be expected to have an adverse affect on the clinical interpretation.

d. Precision

Three pooled samples were analyzed for grouped intra-assay reproducibility. The same samples were assayed as duplicates during fifteen separate runs, Table 20.

2. Thyroxine

a. Sample Preparation

Approximately 200 $\mu\ell$ of serum is placed in a sample cup which is automatically diluted 32-fold in a buffer containing 8-anilino-1-napthalene sulfonic acid (ANS)

Table 16
CORTISOL STANDARD CURVE

A. Actual Data

#	Concentration (ng/ml)	%Bound	Counts
1	0.000	64.1	9,615
2	2.000	60.0	9,772
3	4.000	56.6	9,972
4	8.000	50.8	9,827
5	16.000	40.7	9,845
6	32.000	32.5	9,811
7	64.000	24.7	10,066

B. Standards Estimated on Self Curve

Concentration (ng/ml)		Difference
Actual	Calculated	(%)
0.00	0.177	*****
2.00	2.000	0.0
4.00	3.725	6.9
8.00	7.370	7.9
16.00	17.14	7.1
32.00	32.07	0.2
64.00	63.62	0.6

C. Curve Parameters

Mean Total Count	9866
CV	1.46
Slope	−1.0
Intercept	3.043
Bo	64.6%
Bm	11.6%
Confidence Factor	0.95

Table 17
CORTISOL CLINICAL CORRELATION

X-Axis	Y-Axis	N	Coefficient	Slope	Y-Intercept
ARIA II®	Kit I	29	0.99	0.985	0.62
ARIA II®	Kit II	27	0.97	0.879	0.04

Table 18
CORTISOL PROPORTIONAL DILUTION STUDY

%Pool	%Diluent	Dilution factor	N	Experimental value (ug/dl)	SD	%CV	Corrected value (ug/dl)
80	20	1.25	9	43.44	2.32	5.3	54.3
60	40	1.67	9	34.79	1.22	3.5	58.10
40	60	2.5	9	22.64	1.45	6.4	55.75
20	80	5.0	9	10.44	0.86	8.3	52.2

Table 19
CORTISOL SPECIFICITY

Compound	% Cross reactivity[a]
Cortisol	100.0
Corticosterone	28.4
Cortisone	26.4
6-β-OH-Cortisol	28.0
Prednisone	38.9
Prednisolone	66.0
11-Deoxycortisol	4.9
11-Deoxycorticos-terone	1.5
Dexamethasone	3.2
Aldosterone	1.5
Testosterone	0.01
Dihydrotestoster-one	0.01
Progesterone	0.01
17-α-OH Proges-terone	0.01
Androstenedione	0.01
Dehydroepian-drosterone	0.01
17-β-Estradiol	0.01
Estriol	0.01

[a] At 50% of Bo.

Table 20
CORTISOL REPRODUCIBILITY

Parameter	Sample	Low pool	Medium pool	High pool
		A. Intra-Assay (Grouped)		
Mean (ug/dℓ)		3.7	14.4	36.7
SD		0.4	0.8	2.1
CV		11.0	5.5	5.7
N		10.0	10.0	10.0
		B. Inter-Assay (Duplicates)		
Mean		4.2	15.6	37.8
SD		0.4	1.5	3.0
CV		9.0	9.9	7.9
N		15.0	15.0	15.0

which displaces thyroxine from TBG. For samples where volume is limited a manual dilution of 20 $\mu\ell$ of sample and 500 $\mu\ell$ of sample buffer may be assayed on ARIA II®. However, a different floppy disc is required for each method.

b. Standard Curve

A typical standard curve for thyroxine between 1 and 32 ug/dℓ is provided in Table 21. The 0.97 confidence factor implies excellent standard curve fit.

c. Accuracy

Clinical correlation — Clinical samples were assayed by a competitive protein bind-

Table 21
THYROXINE STANDARD CURVES

A. Actual Data

#	Concentration (ug/dl)	% Bound	Counts
1	0.000	45.5	14,590
2	1.000	40.6	14,767
3	2.000	36.2	14,690
4	4.000	30.3	14,465
5	8.000	22.5	14,612
6	16.00	16.3	14,572
7	32.00	11.3	14,168

B. Calculated Results on Self Curve

Concentration		Difference
Actual	Calculated	(%)
0.00	0.062	*****
1.00	0.949	5.1
2.00	1.985	0.8
4.00	3.934	1.7
8.00	8.392	4.9
16.0	16.03	0.2
32.0	31.62	1.2

C. Standard Curve Parameters

Parameter	
Mean total count	14.527
CV	1.33
Slope	−1.0
Intercept	1.889
Bo	45.9
Bm	4.0
Confidence Factor	0.97

Table 22
THYROXINE CLINICAL CORRELATION

X	Y	Slope	Intercept	Correlation coefficient	N
ARIA II®	Method I	1.02	−0.32	0.99	20
ARIA II®	Method II	0.98	0.22	0.95	28

diluted proportionately with T_4—free serum up to five-fold. The CV's for 10 replicate determinations ranged between 5.2 and 9.1, Table 23. The overall corrected value mean for the dilutions was 15.8 with a CV of 1.6 indicating excellent linearity on dilution.

Proportional dilution study — A serum pool containing 16 ug/dl thyroxine was diluted proportionately with T_4—free serum up to five + fold. The CV's for 10 replicate determinations ranged between 5.2 and 9.1, Table 23. The overall corrected value mean for the dilutions was 15.8 with a CV of 1.6 indicating excellent linearity on dilution.

Table 23
THYROXINE PROPORTIONAL DILUTION STUDY

%Pool	%T4-Free serum	Dilution factor	Experimental values (ug/dl)	SD[1]	%CV	Corrected values
100	0	1.0	16.15	1.16	7.2	16.2
80	20	1.25	12.83	0.67	5.2	16.0
60	40	1.67	9.51	0.63	6.7	15.9
40	60	2.5	6.28	0.57	9.1	15.7
20	80	5.0	3.10	0.22	7.1	15.2

Table 24
THYROXINE SPECIFICITY

Compound	Cross reaction[a]
L-T_4	100.0%
D-T_4	100.0%
L-T_3	0.17%
D-T_3	0.26%
Salicylic acid	0.01%
Phenylbutazone	0.01%
Diphenylhydantoin	0.01%
Iodotyrosine	0.01%
Diiodotyrosine	0.01%

[a] At 50% of Bo.

Specificity — The thyroxine antibody produced in rabbits with thyroxine conjugated to bovine thyroglobulin through a 2,4-toluene diisocyanate linkage cross reacted equivalently with D-thyroxine. Triiodothyronine had less than 0.3% cross reaction. Other iodinated tyrosines had no effect, Table 24.

d. Precision

Intra- and inter-assay reproducibility studies were performed on three pools by the automatic dilution method, Table 25. The CV's were less than 8 for the 20 inter-assay runs of duplicates and less than 6 for the intra-assay run.

3. T_3 Uptake

a. Sample Preparation

About 200 μl of serum is placed in a sample cup and 16 μl of the sample is diluted with ^{125}I-T3 tracer 31-fold automatically. Since the equilibrium takes place rapidly between T_3 and the serum protein, the free T_3 is separated from the serum proteins by an antibody chamber, all within 2.3 min.

b. T_3 Uptake Calibration

A pooled normal reference serum is provided for calibration purposes for use on ARIA II®. This reference sample is calibrated against a pooled primary reference solution to maintain T_3 Uptake calibration. The reference pool has an Uptake of 30% for the mid-point of normals. The percent uptake for unknowns is calculated according to Equation 11.

$$T_3 \text{ Uptake} = \frac{(\%B\ (S))\ (\%\text{Uptake}\ (R))}{\%B\ (R)} \tag{11}$$

Table 25
THYROXINE REPRODUCIBILITY DATA

Parameter	Low pool	Medium pool	High pool
		A. Intra-Assay	
Mean	2.7	8.2	16.5
SD	0.14	0.34	0.91
CV	5.2	4.2	5.5
N	20.0	20.0	20.0
		B. Inter-Assay[a]	
Mean	2.7	8.1	16.4
SD	0.20	0.61	0.77
CV	7.6	7.5	4.7
N	20.0	20.0	20.0

[a] Duplicate on each run.

where %B(s) is the percent binding of the sample, %Uptake (R) is the percent binding of the reference. Since the primary reference has a %B of 30, the T_3 Ratio is calculated by dividing the %Uptake by 30.

C. Precision

The precision of the method is related to the reproducibility of the percent binding directly, without relating to a standard curve. Intra- and inter-assay reproducibility data for low, normal and high pools are presented in Table 26. There are numerous factors which affect T_3 Uptake data, but in combination with thyroxine data an approach to free T_4 concentrations in serum is often utilized for assessing thyroid status.

4. Digoxin

a. Double First Antibody Method

Sufficient digoxin specific antibody is placed in antibody chambers to bind more than 85% of the digoxin tracer (^{125}I tyrosyl-3-0-digoxin). The concentration of free antibody is adjusted to bind about 65 to 80% of digoxin as observed with the antibody chamber. The time for equilibration of digoxin tracer with free antibody was dependent on the concentration of digoxin, Figure 9. A 2 ng/mℓ standard reached equilibration in 10 min, while the tracer alone took 30 min. Digitoxin, a potential competitive inhibitor was found to interfere with tracer binding to the free antibody, as shown by a drop in cross reactivity from 17.5% to less than 1% as the incubation time increased from 8 min to 60 min, Figure 10. A compromise on digitoxin cross reactivity and minimum time for good reproducibility was determined to be 18 min.

The effect of ambient temperature was studied on the incubation of digoxin with free antibody. With three instruments operating at either 18 or 29 ± 1°C there was no significant effect on overall performance, Table 27, as shown by performance of three groups of low, medium, and high pools in three positions of the carousel. Ambient temperature changes greater than 4°C during a run will affect the apparent values.

b. Digoxin Standard Curve

The DFA digoxin assay allows two methods of running standards and unknowns. The standards are placed in sample cup positions 1 to 6 and controls and unknowns may be placed in positions 7 to 58 in singlet or duplicate cups. The standards and

Table 26
T_3 UPTAKE REPRODUCIBILITY

Parameter	Hypethroid	Euthyroid	Hypothyroid
	A. Intra-Assay		
Mean (%Uptake)	57.0	31.0	17.2
SD	1.1	1.0	0.7
CV	2.0	3.4	4.2
N	10.0	10.0	10.0
	B. Inter-Assay		
Mean (%Uptake)	66.5	32.3	18.4
SD	4.1	1.5	1.0
CV	6.2	4.8	5.4
N	15.0	15.0	15.0

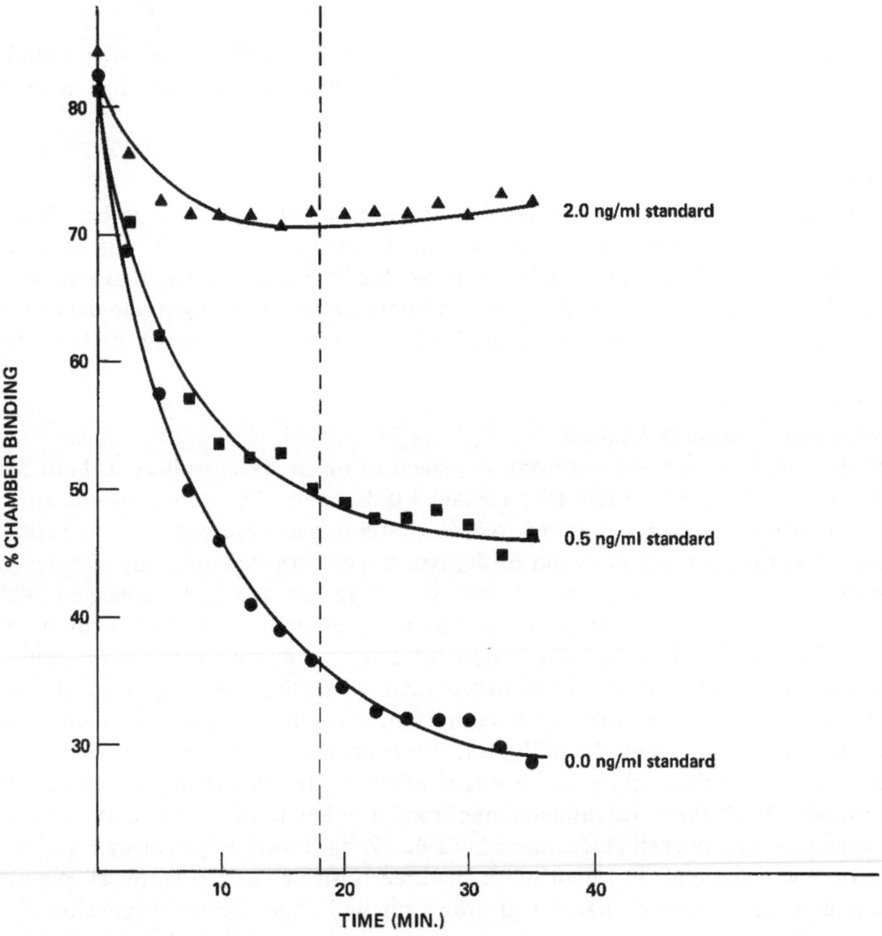

FIGURE 9. Digoxin equilibration time with free antibody. Digoxin standards were incubated in sample cups with free digoxin specific antibody and tracer for the times indicated and then separated with an antibody chamber.

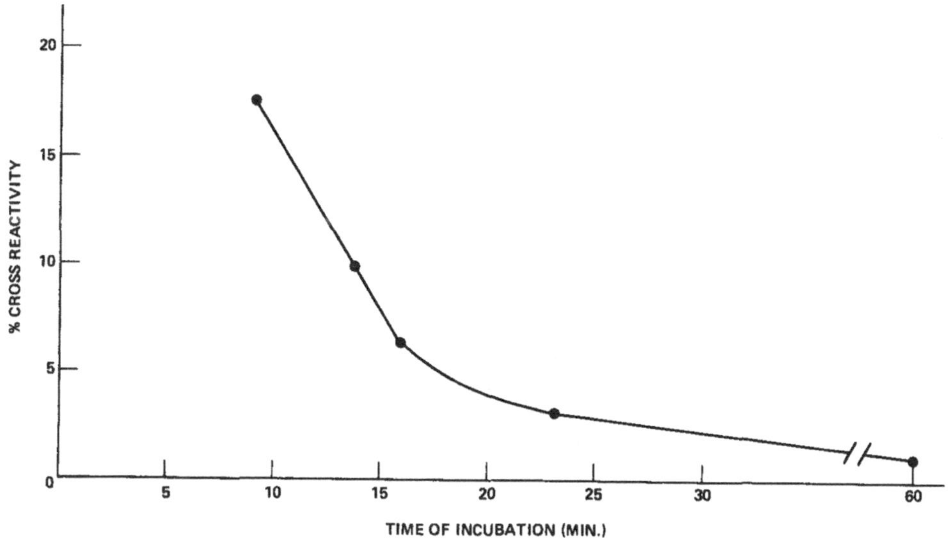

FIGURE 10. Digitoxin cross reactivity with time of incubation. Typical high levels of digitoxin were incubated in a typical double first antibody assay for digoxin.

Table 27
LACK OF AMBIENT TEMPERATURE EFFECT ON DIGOXIN ASSAYS

	Ambient Temperature (°C)					
	29°C			18°C		
Low pool (N = 18)[a] carousel position	7—9	25—27	43—45	7—9	25—27	43—45
ng/mℓ	0.86	0.87	0.92	0.83	0.83	0.83
CV	4.0	8.4	5.2	6.2	5.4	6.1
Medium pool (N = 18) carousel position	10—12	28—30	46—48	10—12	28—30	46—48
ng/mℓ	2.07	2.14	2.12	2.00	2.04	1.99
CV	3.9	5.6	6.0	5.4	5.1	7.3
High pool (N = 18) carousel position	13—15	31—33	49—51	13—15	31—33	49—51
ng/mℓ	3.90[b]	3.94	3.84	3.68	3.70	3.68
CV	6.6	6.9	8.4	9.3	4.7	10.6

[a] Triplicate samples were run as groups of three on three instruments with two cycles per sample cup. The means and CVs were determined for each location in the carousel.

[b] N = 15 for this set.

unknown samples are placed in the inner circle of the carousel while empty cups are placed in the corresponding positions in the outer circle of the carousel for the incubations to be set up with free antibody. Duplicates can be assayed from two sequential sample cups with one pass through the carousel or by placing twice as much sample in the sample cup (0.150 $\mu\ell$). Each standard and unknown may be assayed twice from the same sample cup and incubation cup by repeating the run automatically with two passes through each cup.

Table 28
DIGOXIN STANDARD CURVE DATA

A. Actual Data

#	Concentration (ng/mℓ)	% Bound 1	% Bound 2	Counts 1	Counts 2
1	0.0	73.1	72.2	8016	8193
2	0.5	60.3	59.7	8396	8051
3	1.0	47.4	48.2	8529	8075
4	2.0	34.3	34.7	8033	8250
5	4.0	22.4	22.6	8570	8346
6	8.0	15.3	15.7	8370	8132

B. Standards Estimated on Self Curve

	Concentration (ng/mℓ)		Difference%	
	Calculated			
Actual	1	2	1	2
0.0	−0.027	−0.023	***	***
0.5	0.506	0.519	1.3	3.7
1.0	1.057	1.024	5.7	2.4
2.0	1.950	1.950	2.5	2.5
4.0	3.874	3.959	3.1	1.0
8.0	8.228	8.103	2.8	1.3

C. Curve Parameters

	Curve 1	Curve 2
Mean total count	8313.0	8187.0
CV of total count	2.89	1.37
Slope	−1.40	−1.40
Intercept	0.440	0.511
Bo	72.4%	71.4
Bm	10.5%	10.8
Confidence factor	0.97	0.97

The standard curve data in Table 28 includes data from duplicate results from two typical passes through the carousel. The percent binding of digoxin to the free antibody is represented which is simply 100-%B to the antibody chamber. The first and second results from the same sample cup are provided. Between the two standardization, 54 cycles were run. As the antibody loses binding capacity with extended use, the apparent change in binding of the free antibody increases. The largest effect is apparent at high antigen concentrations, e.g., 8 ng/mℓ. A typical chamber changed from 74.9 to 75.8 and 14.4 to 19.6 %B to free antibody for the 0 and 8 ng/mℓ standards, respectively, with over 600 cycles of unknowns and standards. By running single cup duplicates the variability is reduced because of the reduction in perturbation of the duplicate standardization.

c. Accuracy

Clinical correlations with manual kits — Over a hundred clinical samples containing digoxin were assayed by ARIA II® and three manual kits. Kit I was the Becton Dick-

Table 29
DIGOXIN CLINICAL COMPARISONS

X-Axis	Y-Axis	N	Correlation coefficient	Slope	Y-Intercept
ARIA II®	Kit I	109	0.98	0.92	0.01
ARIA II®	Kit II	112	0.92	0.94	0.26
ARIA II®	Kit III	114	0.95	0.93	0.04

Table 30
DIGOXIN LINEARITY STUDY

Clinical sample	Dilution factor	Observed ng/mℓ	Dilution corrected	Percent observed
A	1	2.40	2.40	100
	2	1.13	2.26	94
	4	0.59	2.36	98
B	1	2.18	2.18	100
	2	1.15	2.30	106
	4	0.57	2.28	105
C	1	5.95	5.95	100
	2	2.62	5.24	88
	4	1.34	5.36	90
	8	0.67	5.36	90
D	1	2.42	2.42	100
	2	1.26	2.52	104
	4	0.63	2.52	104

inson Immunodiagnostics Liquid Phase ^{125}I-Digoxin, Kit II was a Commercial Liquid Phase ^{125}I-Digoxin method, and Kit III was a solid phase ^{125}I-Digoxin method. From regression analysis, Table 29, all kits had a correlation coefficient greater than 0.92. Generally there was a slight positive bias on ARIA II®.

Recovery and linearity — Digoxin (1./0 ng/mℓ) was added to ten clinical samples containing 0.3 to 3.0 ng/mℓ digoxin. The mean recovery was 104 ± 14.0%. Four clinical samples were diluted serially and analyzed for digoxin, Table 30. By setting the undiluted sample at 100% of expected, the corrected values ranged from 88 to 106% of expected. Clinical sample C showed more consistency on dilution, indicating the undiluted sample may have assayed slightly high. In either case the sample was clearly in the therapeutic range.

Specificity — Lanatoside C and Desacetyl Lanatoside C were found to cross react with digoxin as would be expected since the only differences are in the carbohydrate side chain. Digitoxin and dihydrodigoxin cross react with the free antibody at 2.1 and 1.6%, respectively, Table 31.

d. Precision

Intra-assay reproducibility of three pools randomly placed in the carousel indicates excellent reproducibility with CV's of 6 or under, Table 32. The CV's are lowered for the samples by taking advantage of two calibrations during the run.

Inter-assay reproducibility was investigated with the same pooled samples on eighteen runs. The samples were run as sequential duplicates and also calculated as single cup duplicates. In either case the results were very comparable with CV's less than 7% for all samples, Table 33.

Table 31
DIGOXIN SPECIFICITY

Compound	%Cross reactivity
Digoxin	100%
Digitoxin	2.1%
Lanatoside C	75%
Desacetyl lanatoside C	83%
Dihydrodigoxin	1.6%
Gitoxin	1.0%
Strophanthidin	1.0%
Cortisol	0.1%
Oubain	0.1%
Estradiol	0.1%
Spironolactone	0.1%
Cholesterol	0.1%
Progesterone	0.1%
Testosterone	0.1%

Table 32
DIGOXIN INTRA-ASSAY REPRODUCIBILITY — RANDOM

	Digoxin	
	$\bar{X}$, ng/mℓ	CV
Low pool (N = 10)		
1st Carousel	0.89	3.4
2nd Carousel	0.85	4.9
Single cup duplicate	0.87	3.2
Medium pool (N = 8)		
1st Carousel	2.07	5.4
2nd Carousel	2.07	4.6
Single cup duplicate	2.07	2.1
High pool (N = 10)		
1st Carousel	4.43	6.2
2nd Carousel	4.47	5.6
Single cup duplicate	4.45	4.5

Carry-over effect on precision was investigated by running 0.5 and 4.0 ng/mℓ standards in groups of ten or alternately as duplicates. There was no apparent carry-over affect of a 4.0 preceeding a 0.5 ng/mℓ or a 0.5 preceeding a 4.0 ng/mℓ, Table 34.

V. SUMMARY

Versatility in automated RIA has been achieved with the ARIA II® system. Several different types of RIA methods may be used which take advantage of a reusable antibody chamber. In many cases, the turn around time from placing a sample on the instrument to result can be achieved in less than 3 min if the instrument is standardized and running the particular assay. This rapid turnaround from sample aspiration to result permits appropriate quality control checks to be performed before the run is committed on valuable unknown samples.

Table 33
DIGOXIN INTER-ASSAY REPRODUCIBILITY

	Digoxin	
	$\overline{X}$, ng/mℓ	CV
Low pool (N = 18)		
1st Carousel	0.88	5.9
2nd Carousel	0.85	5.7
1st Single cup	0.87	5.2
2nd Single cup	0.86	5.2
Medium pool (N = 18)		
1st Carousel	2.08	4.8
2nd Carousel	2.03	4.2
1st Single cup	2.05	3.9
2nd Single cup	2.06	4.6
High pool (N = 18)		
1st Carousel	4.55	4.7
2nd Carousel	4.48	6.7
1st Single cup	4.51	3.4
2nd Single cup	4.53	6.4

Table 34
DIGOXIN CARRY-OVER

Condition evaluated 0.5 ng/mℓ	Mean (ng/mℓ)	SD	CV	N
Sequential	0.47	0.03	5.1	10
Alternate	0.55	0.03	5.2	10
Single sample following 4.0 ng/mℓ	0.55	0.04	6.7	5
Single sample following 0.5 ng/mℓ	0.55	0.02	3.8	5
4.0 ng/mℓ				
Sequential	3.82	0.15	4.0	10
Alternate	3.86	0.16	4.2	10
Single sample following 4.0 ng/mℓ	3.84	0.17	4.4	5
Single sample following 0.5 ng/mℓ	3.81	0.15	4.1	5

The use of a flexible disc to program the instrument allows the introduction of new procedures with differing characteristics to be readily adapted to automation. The use of double first antibody techniques has opened up new areas where sensitivity is a requirement.

REFERENCES

1. **Berson, S. A. and Yalow, R. S.** Isotopic tracers in the study of diabetes, in *Advances in Biological and Medical Physics*, Academic Press, New York, 1958, 349.
2. **Johnson, L. R.**,U.S. Patent 3,896,217, 1975.
3. **Johnson, L. R.**, U.S. Patent 4,059,685, 1977.
4. **Reese, M. G.**,U.S. Patent pending.
5. **Reese, M. G.**,U.S. Patent 4,108,976, 1978.
6. **Eldridge, J. S. and Crowther, P.**, Absolute determination of I^{125} in clinical applications, *Nucleonics*, 22, 57, 1964.
7. **Watts, P. and Raty, D.**, Data Reduction for ARIA II® Radioimmunoassay, Becton Dickinson Immunodiagnostics, 1978.
8. **Watts, P. and Raty, D.**, Curve Fitting with Data that is not "Normally" Distributed, Becton Dickinson Immunodiagnostics, 1978.
9. **Lindberg, B. S., Johansson, E. D. B., and Nilsson, B. A.**, Plasma levels of unconjugated oestrone, oestradiol-17, and oestriol during uncomplicated pregnancy, *Acta Obstet. Gynecol. Scand., Suppl.*, 32, 21, 1974.
10. **de Hertogh, R., Thomas, K., Bietlot, Y., Vanderheyden, I., and Ferin, J.**, Plasma levels of unconjugated estrone, estradiol, and estriol, and of HCS throughout pregnancy in normal women, *J. Clin. Endocrinol. Metab.*, 40, 93, 1975.
11. **Loriaux, D. L., Ruder, H. J., Knab, D. R., and Lipsett, M. D.**, Estrone sulfate, estrone, estradiol and estriol plasma levels in human pregnancy, *J. Clin. Endocrinol. Metab.*, 35, 887, 1972.
12. **Stafford, J. E. H. and Watson, D.**, Total and unconjugated serum oestriol levels as a test of foetoplacental function, in *Radioimmunoassay and Related Procedures in Medicine*, Vol. II, International Atomic Energy Agency, Vienna, 1974, 57.
13. **Rolleri, E., Zannino, M., Orlandini, S., and Malvano, R.**, Direct radioimmunoassay of plasma cortisol, *Clin. Chim. Acta*, 66, 319, 1976.

Chapter 4

TECHNICON AUTOMATED RIA SYSTEM: STAR®

S. L. Ossad, F. Savino, and E. Cohen

TABLE OF CONTENTS

I. INTRODUCTION

A. Growth

In the past two decades, radioimmunoassay (RIA) has evolved from a highly demanding research procedure into a versatile and widely accepted diagnostic tool. The present annual volume of tests, estimated at 150 million, attests to its acceptance. This rapid growth reflects, in great part, the enormous progress made in the development of necessary reagents and new approaches to the problems that have until now blocked true automation.

B. Requirements of RIA

Early assays routinely involved sample pretreatment, lot dependent adjustments to the tracer, short useful life of tracer, incubations of several days duration, counting times of several minutes, three or more replicates per sample, and extensive manual data manipulation. Withal, the precision was only just acceptable. Many of these limitations have been overcome by advances in reagent preparation.

Progress in tracer manufacture has been significant:

1. Iodine^{125} (^{125}I) has replaced Iodine^{131} as the gamma tracer of choice, yielding a longer useful life. Further, newly developed carrier free ^{125}I permits a higher degree of incorporation of the signal producing nuclide into the target molecule.
2. Improvement of the manufacturing process of the radionuclide ^{125}I has eliminated deleterious ancillary products, further improving tracer stability.
3. Derivatization of steroids and nonpeptide molecules has circumvented the use of β-emitters such as tritium and carbon-14. Such molecules are now routinely tagged with ^{125}I, simplifying sample handling as well as reducing per sample reagent cost.
4. The radioiodination procedure itself has been rationalized; the amounts of Chloramine-T have been cut significantly, lessening oxidation damage to the biomolecules. Alternate approaches to Chloramine-T, such as lactoperoxidase and coupling of a radio-iodinated ligand to the hapten, permit a selection of a most favorable approach.

The overall result has been the routine preparation of radio-iodinated compounds with good radioimmunoreactivity, adequate stability to permit routine commercial distribution, and a higher specific activity enabling a high throughput at the scintillation counter.

There has been a parallel development in the antibodies used in RIA:

1. Elaborating on the original procedure for developing antibodies to nonimmunogenic haptens,[1] it has become possible to produce antisera of high specificity by controlling the site of coupling of hapten to carrier. This, in turn, has often eliminated the need to isolate the analyte of interest from the sample matrix. Such antibodies, combined with displacing agents have superceded the original competitive protein binding assays which exploit endogenous serum proteins. Once more, extraction or denaturation of binding proteins has been eliminated, thus simplifying assays and improving their precision.
2. Purification of the immunogens has reduced cross-reactivities in some cases by elimination of antibodies produced to the contaminants.
3. Improved immunization procedures have improved the yield of suitable material, often while utilizing smaller quantities of the immunogen. Higher affinity prod-

ucts in larger quantities have reduced the length of incubations required and lowered the costs of antisera as well.

The sum of these advances is the commercial availability of quality reagents and simplified protocols, offering more reliable results while using smaller samples.

In spite of these developments, RIA has remained a labor intensive discipline requiring in general a high level of skill and commitment. The equipment that has evolved for the RIA lab has eased some of these problems, but has not eliminated them.

Accurate manual pipettors render the dispensing of microliter volumes routine but require tip-wiping and precision may suffer from pipet fatigue over the length of a day. More sophisticated pipetting stations have been designed for high volume facilities, but are not generally available where routine RIA work is performed. Their initial cost, plus the required set-up and take-down times make them unattractive for most RIA labs. Developments in detection devices and data handling equipment have been dramatic but have not accounted for significant overall automation of RIA. The semi-automated systems still require that the technologist present each sample to the dispenser and then present a tube into which the reaction mixture is placed.

Coated tube technology and centrifuges adapted to the array of small tubes used in RIA have simplified the pivotal separation step. Nevertheless, each tube once again must be handled several times as it is moved through this step and then loaded into a scintillation counter.

Even if the design of the assays will permit all these manipulations without intolerable loss of precision and accuracy, the technician's involvement is far greater than is acceptable for an automated system designed for routine use.

The newer generation of reagents now make it possible to reconsider the automation of the total RIA procedure: the reagents are consistent, more stable, and require fewer steps to accomplish the assays. Basically, the protocol now consists of the precise addition of two to four reagents, waiting an interval of time, separating the reaction products, counting one of the separated reaction components and then data reduction.

C. Technicon

It is logical that Technicon would involve itself in the development of an automated RIA system since it is the leader in the field of laboratory automation. Technicon's primary contribution to automated instrumentation and the basis for its continuing success is the concept and application of continuous flow (CF) technology. Though it utilizes a few simple conditions, CF offers enormous versatility and has been applied to many clinical and industrial instruments to date.

In its simplest form, CF is embodied in the precise flow of an air segmented liquid through a tube.[2] The constant flow rate is achieved by a precision peristaltic pump which drives fluid through calibrated pump tubes. A selection of pump tube diameters allows a variety of flow rates. The pump used on Star®, for example, can accommodate ten fluid lines and four air lines. Thus it is a proportioning pump capable of delivering a number of liquids at different rates into a stream (known as the "analytical stream"). The four air lines permit the segmenting of different analytical streams (Figure 1).

The segmenting air bubbles are key to the actualization of continuous flow in a system. They isolate the fluid segments, known as slugs. Each slug is actually an isolated reaction mixture. The occluding air bubble between slugs acts as a barrier to dispersion, or the carryover of the contents of one slug into the next. In fact, the barrier is not absolute and its effectiveness depends upon the experimental conditions. Dispersion in CF is similar in nature to the dispersion encountered in chromatography and counter current distribution. By controlling the experimental conditions, including

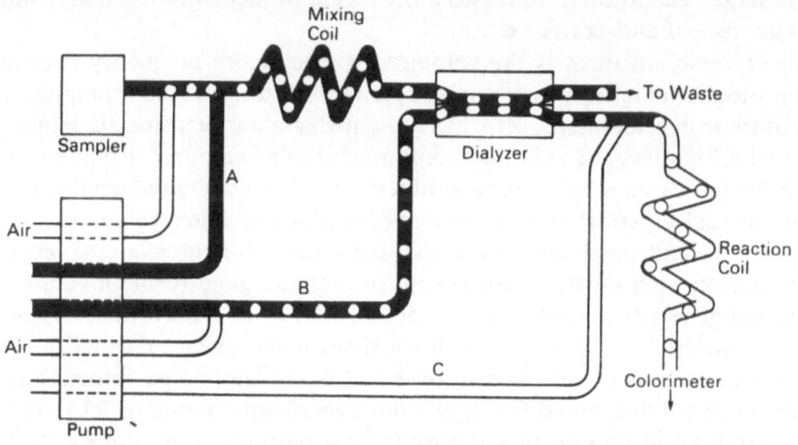

FIGURE 1. Example of continuous-flow analysis scheme. A, B — Diluents. C — Reagent stream.

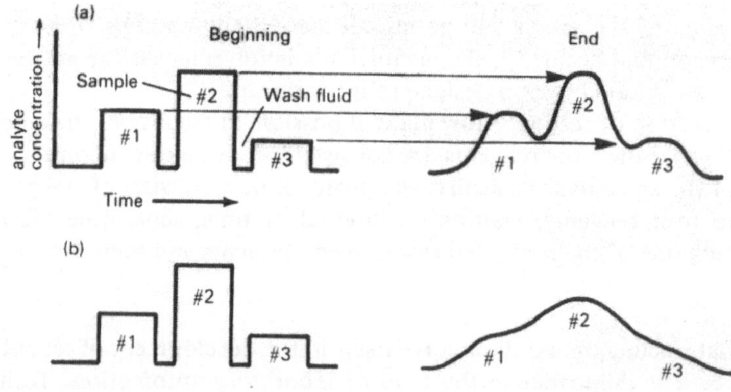

FIGURE 2. Effect of dispersion during continuous-flow analysis on sample concentration curve shape. A. Acceptable dispersion; analyte and analyte-chromogen concentrations plotted on molar basis. B. Unacceptable dispersion.

flow rate, reagent formulation, and bubble size, dispersion or carryover can be kept within acceptable limits.

The air bubble serves as a "sweep" as well, pushing the slug before it. To be effective, it must occlude the tube satisfactorily. Figure 2 illustrates this. The effectiveness of the air bubble as a "sweep" permits the pumping of slurries through the tubes (as required in Star®).

Thorough and consistent mixing is essential for reliable results. Mixing within each slug occurs as a natural consequence of the slugs' movement through the tubing. This is known as "bolus mixing" and is illustrated in Figure 3. Thus continuous flow has the important advantage of achieving excellent mixing without recourse to external means.

A requisite for maintaining a constant flow rate is an unchanging volume. Therefore, whenever sample or any reagent is not being introduced into the analytical

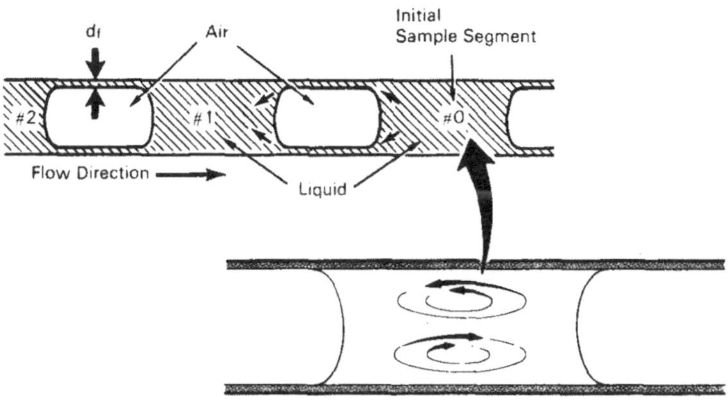

FIGURE 3. A. Sample dispersion during segmented flow through open tube with marker-dye sample injected into segment #0. Film of liquid phase (d_f = film thickness) from segment #0 will be overtaken and mixed into segment #1, allowing dye to move from #0 to #1; dye built up within segment #1 can also be transferred to #2. Dye will become dispersed over many following segments. B. Bolus circulation within moving liquid segment occurs naturally in any segmented flow system.

stream, an equal volume of inert reagent — buffer or wash solution — must be alternated with it. In practice, for example, samples are alternated with buffer. Each sample occupies a certain number of slugs, with the intervening slugs carrying buffer instead of sample. These intervening slugs serve to wash off any sample that may cling to the tubing wall as the reaction stream travels past it. Since the flow rate is constant and well controlled, sampling and reagent switching may be controlled by timers. For example, Star® uses a 60 sec cycle; that is, 60 sec elapse between the beginning of a sample pick-up and the beginning of the next sample pick-up. The sample is injected into an analytical stream that has an occluding bubble every 2 sec, yielding 30 slugs/min. Sample enters 18 of these slugs, meaning sampling is for 36 sec. The other 12 slugs carry buffer and serve as intersample wash. The important consideration is that all samples, standards or calibrators are timed identically and travel the identical path under identical conditions.

Technicon's CF instrumentation is of a modular nature. Most designs are a reintegration of existing components with modification where necessary, into a final system. The components have usually included sampler, mixing coils, reaction coils, dialyzers, colorimeters, and, of necessity, pumps. When combined into a particular configuration, they have performed the desired analyses.

Having reviewed continuous flow, it may be useful to look at its advantages and constraints.

The advantages:

1. The only sample variable consumable is the sample cup. The other consumables, e.g., pump tubes and dialysis units, are system fixed and replaced after a preset number of hours of use.
2. Each sample and standard or calibrator goes through the identical path under identical conditions.

3. Reagent volumes are small, reagent consumption and cost is low.
4. CF is extremely flexible, offering the possibility of automating virtually all methods of interest.
5. Effective mixing is achieved without external means.

The constraints:

1. A particular flow stream, or channel as it is called, is dedicated to a single analysis. Multiple analyses require multiple channels. Multiple channels are to be found on SMAC (20 channels) and SMA instruments (12 to 18 channels).
2. While proportioning is precisely controlled, CF cannot be used to dispense absolute quantities. Each pump tube may be relied upon to deliver a constant volume, but that volume may vary by ±10% of the nominal rating.
3. There is a practical limit to the length of the flow stream. As noted earlier, the occluding bubble is not an absolute barrier; over a long enough path length, dispersion becomes measureable.
4. A constant volume must be maintained in the flow stream. Therefore, whenever a reaction component is not entering the stream, another liquid must make up the volume.

II. STAR® CONCEPT

With this brief introduction, we can now consider Star® in particular. The unique feature of the Star® RIA system is its means of separation of bound from unbound antigen at the reaction's end. This separation step is crucial to the success of any radioimmunoassay. Addition of separating media such as charcoal and other adsorbents create, in fact, a secondary equilibrium between the complex and the adsorbent. To minimize this effect, contact time is controlled and minimized and sometimes this step is carried out at lowered temperatures. Precipitation by ammonium sulphate, polyethylene glycol, or via second antibody carries the potential error of bringing free antigen down in the precipitating matrix. Antibody coated tubes eliminate these pitfalls. However, there can be great tube-to-tube variability of antibody concentration and residual-free antigen may adhere to the tube.

A. Magnetic Separation

In Star®, separation is effected by the use of a solid phase antibody wherein the antibodies are coupled to magnetically susceptible particles. These particles, in the 1 to 10 μm range, consist of submicron ferric oxide particles entrapped in an organic polymer matrix. Antibodies are then covalently linked to these particles. The ferric oxide: polymer ratio is controlled to permit satisfactory susceptibility to magnets while still showing acceptable flow characteristics when pumped as a slurry.

The antibody particles in a slurry are titered and evaluated exactly as antisera would be. The final concentration of particles is adjusted with immunologically inactive particles to a final concentration of about 20 mg/mℓ. This slurry will trigger an optical sensor by occluding a light passing through the flow stream at right angles to it. The optical sensor is positioned above two electromagnets which straddle the flow stream. The sensor triggers the electromagnets. When on, these magnets hold the magnetizable solid phase against the tubing wall. All antigen immunologically bound to the antibodies are thereby retained at the magnets. Buffer enters the reaction stream between the optical sensor and the magnets. It serves to wash the particles free of unbound antigen. The great majority of the particles are caught at the first magnet, where they are first washed. They are then released back into the stream and travel to the second magnet where they are held back again. This permits an additional wash, particularly of any surfaces occluded at the first magnet. After an interval, the second magnet releases the particles and they travel to the scintillation counter. This then is the heart of Star®: antibody bearing magnetizable particles and the separation of bound from unbound by electromagnets. This separation is gentle and in no way affects the immunologic complex. The antibody particles in slurry have been evaluated just as native antiserum would and used similarly. Thus we are assured of consistent sample-to-sample conditions.

B. Concept to System

Now to turn to the incorporation of this novel idea into a totally automated system.

The Star® instrument consists of an analytical module, shown at the left in Figure 4 and an electronic module to the right. A sampler is placed to the left of the analytical module. It is in the hydraulics of the analytical module that the reagents are mixed, the reaction occurs, the separation is effected and the bound antigen is counted. The electronics module permits the operator to instruct the instrument, control the processes and report the results.

Figure 5 is a schematic of the hydraulic system. It may be simplest to begin at the magnets. The two magnets are represented by the two sets of rectangles which straddle the vertical tube at the right. Just above the magnet housing, a wash buffer line comes into the analytical stream. This delivers buffer that washes the particles held first at the upper magnet and then at the lower one. Just above the buffer inlet is the optical sensor referred to earlier. It is represented by a circle. When the sensor registers the presence of the solid particles, it activates the electromagnets by which the particles are retained to be washed as described.

Below the magnets is a two-position valve. In one position, it allows wash from the magnets to go to waste and simultaneously allows detector wash to clean the coil within the scintillation counter. When the lower magnet releases the washed solid phase, the valve switches to allow the solid phase to travel into the coil within the scintillation counter. The coil's length gives the solid phase and the radioactivity immunologically bound to it a dwell time within the crystal of 45 sec, during which time it is counted. When it passes out of the coil, it goes to the waste receptacle. The counting time per 60 sec cycle is 45 sec as noted. During the balance of 15 sec, an alkaline detergent solution flushes the coil free of any radioactivity material that may have hung back on the tubing wall. Among the monitoring functions is one which alerts the operator if the count rate does not return to background by the time the next sample is about to enter the detector coil.

The optical sensor found above the magnets serves another important function beyond activating the electromagnets; when the sensor registers the presence of solid phase, it starts the clock by which magnets, the scintillation valve, and the counting time of the scintillation counter are controlled. This is done to assure proper phasing, that is, that the sample be at the proper point of the analytical stream at the proper moment. This is necessary in Star® for this system is a departure from Technicon's Continuous Flow technology. Traditionally, Continuous Flow technology is concerned with proportioning only. In Star®, we collect all of a reaction component, the solid phase, and quantitate it at the counter. Thus it is not enough that the ratios within a few of the slugs be correctly determined; the total must be determined reliably. Therefore, all of the solid phase must enter and exit from the counter at the right times. By basing the timing of the counter on the appearance of the solid phase at the electromagnets after it has traveled through most of the analytical stream, we can coordinate

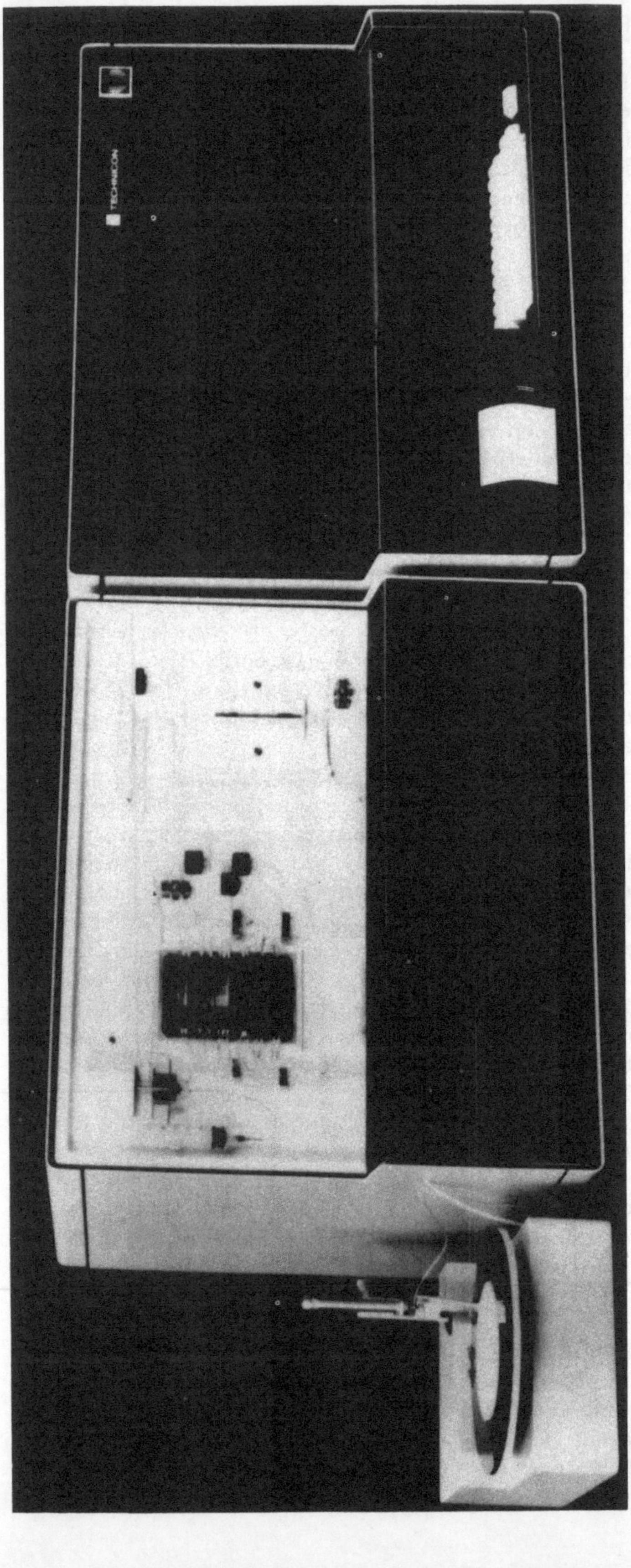

FIGURE 4. Photograph of the Star® System.

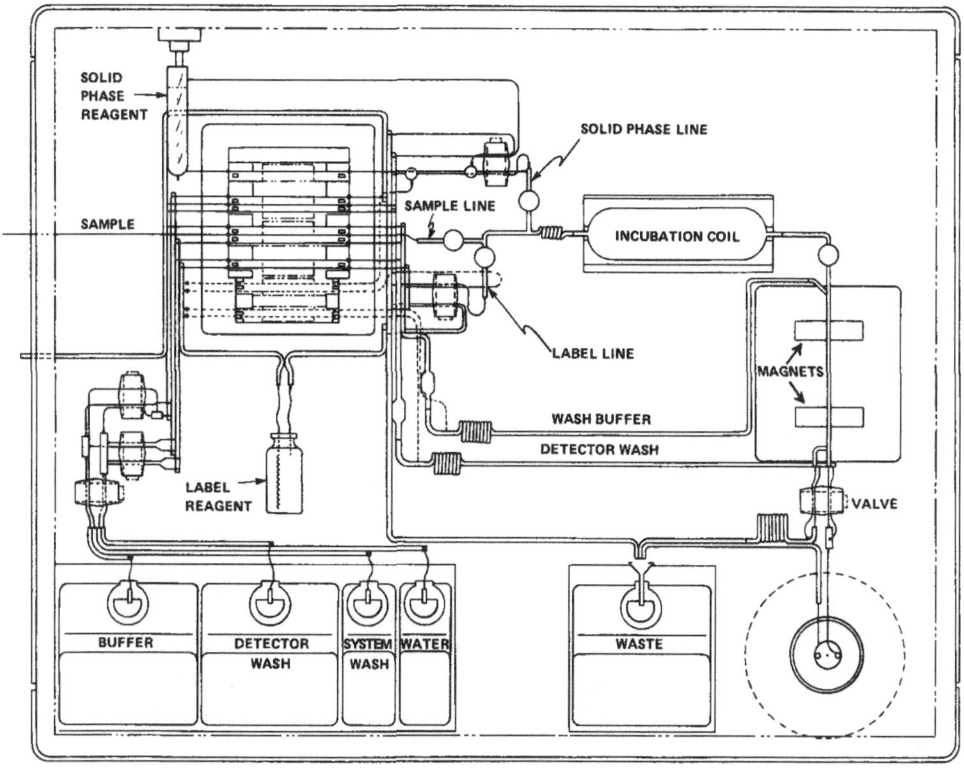

FIGURE 5. Schematic diagram of the Star® hydraulic system.

sample and counter and not worry about any possible surging or other anomaly farther back in the analytical stream. It should be noted that the optical sensor above the magnets is but one of four such devices on the system.

III. COMPONENTS

Star® has a modular design and consists of a sampler, an analytical unit and an electronics unit.

A. The Sampler

The sampler is detachable from the other modules, but is linked via pneumatic and electronic signal and power lines to these units. Its purpose is introduction of samples, standards and controls into the Star® hydraulic system by aspiration from sample cups. A circulating tray of 40 cup capacity is provided. The tray is driven by a 20 r/min motor when 115 V (AC) is applied via electronic signals from the microcomputer. After each rotation pulse is given, the tray advances one cup position and stops by means of a hall effect sensor in the circuit.

Vertical and horizontal probe movement is controlled by the two respective air valves (on the sampler) via air tubing from the solenoid valve panel.

B. The Analytical Unit

The analytical module can be considered to have three discrete functional subsections. In the *Front End*, the solid phase (antibody), label (radioactively tagged antigen), sample, standards, buffers, and wash are proportioned, mixed, reacted, and incubated. After leaving incubation, the reactants pass into the *Separation Assembly* where the unbound label is separated from the bound by magnetic separation. At this point, the material of interest (the bound) passes into the *Counter* assembly where the bound label activity is measured. Finally, all the expended products of the assay are channeled to a waste container equipped with suitable safeguards against possible hazards.

1. Front End

Just as the process flow of Star® moves from left to right (the order of subsections given above), we begin our discussion with the *Front End.*

This subsection consists of the hydraulic assembly and the pneumatic system. The heart of the hydraulic system is the pump module. It is a modified peristaltic pump which has been in existence for a number of years and has been used on other Technicon products. When energized, the pump will rotate at 10 r/min and draw sample, wash, air and reagents through the hydraulic manifold. The pump module is computer controlled in normal operation but may be operated (on-off) via keyboard command.

The mixer assembly is built around a reservoir which maintains solid phase particles in suspension prior to injection into the flow stream. The full assembly consists of the mixing vessel, motor, propeller and shaft. Rotation of the propeller maintains a uniform suspension of solid particles within the mixing vessel. Vessel volume is 25 m*l* and is designed to hold approximately one standard vial of solid phase reagent as supplied in a Star® reagent kit. The vessel itself is coated with a nonwetting agent to prevent any adhesion of the solid phase on the inside surface. The motor, linked to the shaft and propeller, maintains the solution in suspension by rotating at 300 r/min. Control is achieved by a toggle switch located just to the right of the assembly.

Incubation and thorough mixing of the reaction stream (the mixture of solid phase, label, and sample) is accomplished by the incubation coil. This coil is physically placed inside a heating bath.

Coil volume is 7 m*l* and a transit time of 10 min is specified. The construction of the heating bath, containing the incubation coil, allows for direct observation of the mixture during the entire incubation cycle. Temperature is maintained at 37°C with an aluminum heat sink in contact with the coil to assure that the temperature is held constant.

Included in the Star® pneumatic system is a mini-compressor unit, a solenoid valve panel, six pinch valves, and two sampler probe movement valves (discussed earlier). The pneumatic system activates the pinch valves, the probe valves, and supplies air to the manifold in order to create the essential bubble pattern.

Constant air pressure to the solenoid valve panel is supplied by the mini-compressor. The valve panel itself is located at the rear of the analytical module. Controlled by computer signal, air is routed (through tubing) to the specific pinch valve or sample probe valve by the electronic switching of the solenoid valves on the solenoid valve panel.

Six pinch valves determine the flow of the hydraulic streams, selecting reagents in their proper volumes. When activated, the pinch valves allow one set of flow tubing to be pinched; when deactivated, an alternative pathway is closed off. Vertical and horizontal motion is controlled by the two probe valves. A detailed description of each valve follows:

Valve 1 — Controls the supply of buffer, water, wash solution, and detector wash to the system.

Valve 2 — Operates in conjunction with Valve 1 to supply detector wash to the detector wash line. It also operates in conjunction with Valves 1 and 3 to supply water to the main buffer line. Finally, it works together with Valve 1 to supply water or buffer to the sampler wash line.

Valve 3 — Operates together with 1 and 2 to supply water, wash, or buffer to the main buffer line.

Valve 4 — This is the solid phase injection valve and controls the amount of solid phase introduced into the process. Before actual injection of solid phase, the valve will pass buffer through the injection point and pass the solid phase material back to the reservoir via the solid phase return line. During actual injection, it passes the solid phase to the injection point and buffer to the waste line.

Valve 5 — This is the label valve and controls the injection of the label solution into the process. Before injection, the valve passes buffer to the label injection point and label back to the vial via the label return line. During injection, the valve passes label to the label injection point and buffer to the radioactive waste line.

Valve 6 — So called "scintillation valve" because it controls the passage of the analytical stream into the detector coil. In the alternate state of this valve, the incubation coil outflow will be passed to waste and detector wash will be passed to the detector coil.

Valve 7 — The sampler probe valve controlling elevation and pecking by raising or lowering the probe.

Valve 8 — The sampler probe translation control valve which controls the motion of the sampler probe from sample wash to sample cup and from sample cup to sample wash.

After the analytical stream emerges from the incubation coil, it is ready to pass into the Separation Assembly.

2. Separation

The theory of magnetic separation has already been discussed. The key to success, however, rests in the magnetic module. This assembly provides a magnetic field across the tube through which the suspension of solid phase particles and reagents flow. The magnet assembly consists of two magnets, a vibrator, an illuminator bar, and a driver printed circuit broad.

The two magnets are activated by computer signal when the light activated switch (LAS) senses the solid phase particles mixture. The resulting magnetic field attracts the solid phase particles and holds them against the sides of the flow tube. At the same time, the unbound antigen is flushed to waste by a buffer solution which washes over the surface of the captive solid phase particles. The magnets are designed to assure the capture of the maximum possible amount of solid phase and allows a complete wash of unbound antigen.

To assist in the release of solid phase from the magnetic fields on de-energization, a small vibrator is utilized to shake free the solid particles. This unit is simply a solenoid coil and coupling arm. The coupling arm straddles and firmly grasps the flow tube. Based on the LAS timing, the computer will activate the vibrator a short interval after the second magnet is de-energized. Trapping the particles at a second magnet increases recoverability and enables the particles to be fully washed free of unbound tagged antigen.

The illuminator bar located just behind the flow tube lights up when magnetic separation is occurring, thus allowing the operator to view and verify energization.

3. Scintillation Counter

The scintillation detector system consists of a gamma counter and its associated electronics, utilizing a standard sodium iodide detector crystal and Photo Multiplier Tube (PMT) combination.

All electronics for this system, including the high voltage power supply, are housed in the Analytical Module. The detector itself is housed within an inch lead shield to minimize background counts from natural radiation and from isotope reagents in other parts of the instrument.

Radiation, when detected, causes the crystal to emit light pulses. These pulses are detected by the PMT and are converted to pulses of electric current. They are, in turn, fed into a charge sensitive amplifier. Output from this amplifier is filtered and the voltage level is boosted. The amplified voltage signal is then sent to the counter printed circuit board. Pulses that fall within preset levels are counted as real signals known as "event counts". Also counted are signals present which cannot be discriminated (since they occur simultaneously with event counts); they are known as "dead time pulses". Both event counts and dead time pulses are read into the computer for data reduction.

Three energy windows are provided with two preset for ^{125}I and ^{57}Co and a third available for user selection through keyboard commands.

4. Error Detectors

Although the error detectors cannot be considered as a "functional" subsection of the analytical module, they are none the less heavily incorporated in the design of the analytical module. The Star® system employs four identical error detector assemblies installed at different points of the hydraulic flow to sense the absorbance of sample, label and solid phase/incubation outflow. These error detectors perform colorimetric readings using a narrow beam from a red light source (λ = 670 nm) and a detector (silicon photocell) which senses the absorbance of a solution passing through it (see Section III.C.2.e on System Monitoring).

Outputs from the detectors are fed into the Error Detector Board (consisting of four identical circuits) to translate the sensor readings into voltages. Each of the detector channel outputs is multiplexed to an eight bit parallel data line and sent to I/O Board Two and subsequently to the computer. Located on the Error Detector Board as well is the Waste Alarm circuitry which senses the level when probes come in contact with liquid waste in the waste container, thus activating an alarm.

C. Electronics Module

1. Hardware

The third major component of the Star® system is the Electronics Module. It contains the hardware and software that controls all aspects of operation.

a. *Computer*

The brain of the Star® system is the 8080A Intel microprocessor, the most widely known and applied machine of its type now available. The microprocessor, along with its associated circuitry, forms a "computer on a board". This single Printed Circuit Board is referred to here as Central Processing Unit (CPU) Board, more generally as the "computer".

The board is housed in a structure called the card cage. This unit, in addition to holding the CPU board, houses all the boards necessary for the interface of the computer with the 2 disk controllers, the video monitor, the keyboard, memory board and input/output circuits.

The signals traveling between the interfaces and computer share the same lines. This unique orientation of all signals to and from the CPU board is called a BUS. The particular BUS used on Star® is unique to Technicon.

All interfaces, which are incorporated into printed circuit boards (PCBs) may be easily installed or removed from the card cage. Slots are provided in the cage, along with guides for alignment. Insertion or removal of a board is accomplished by simply pushing it into or detaching it from its mating connectors.

Since the computer only executes programs via random access memory (RAM), temporary storage requirements are fulfilled by the memory board, which is directly addressable by the central processing unit (CPU). All program information provided on the system diskette must be transferred to the memory board prior to its execution. This transfer is accomplished automatically during system start-up.

The memory board consists of a number of RAM circuits incorporated on silicon "chips". Up to 16 K bytes of information may be stored on the board at any given time. At the current time, only one memory board is required for Star® system operation.

The input/output (I/O) electronics consists of all the interfaces necessary to link the computer and necessary system devices. The interfaces are required to operate the keyboard, laboratory information system, printer, disk controllers, and I/O boards one and two. The application of each of the above is obvious except perhaps for the two I/O boards.

I/O board one is an interface to the computer to control devices located in the analytical module, such as valves, magnets, vibrator, tray rotation, and pump.

I/O board two is an interface to receive information for the scintillation counter, error detectors (sensors), and waste detector.

b. Video Monitor/Keyboard

These two units are linked and are used for visual output and commands input, respectively. They are both housed in the electronics module and share the same video monitor/keyboard interface board.

The video monitor is a 9 in. (diagonally measured) cathode ray tube (CRT) capable of displaying digital information in the form of a dot matrix. The CRT accepts composite video signals from the computer (via the video/keyboard interface board) and translates these signals into alphanumeric characters. Brightness and contrast controls are incorporated into the front panel for easy operator control.

Significant operator interaction with the Star® system is achieved through the keyboard via circuitry on the video/keyboard interface board. The keyboard consists of 68 keys which are coded for standard alphanumeric keys and special Star® system commands. Activation of the key produces a 7 bit parallel data output to the interface board, which is then passed to the computer.

c. Printer

Hard Copy requirements of the Star® system are met by the printer system, consisting of the printer mechanism and printer interface board. Both are housed in the electronics module.

The core of the mechanism is a dot matrix, impact printing element capable of a rate of 1.25 lines/sec. Sixty-four characters and a number of functional symbols are printed in a format of 40 columns per line.

The interfacing format will accept input data in a parallel bit, serial character format. Characters are encoded by the central processing unit (CPU) and supplied to the printer on eight bit data lines. A roll-type forty columns, 3 7/8 in. pressure sensitive paper is utilized. Multiple copy capability is provided.

d. Flexible Disk System

The flexible disk system of Star® allows for fast and easy storage and retrieval of the Star® operating program. Elements of the system include the disk storage drive, 2 disk drive controllers, and the media on which all the information is stored, the flexible diskette, also known as a "floppy disk".

A diskette may be likened to a 45 r/min record which has been enclosed in an 8 by 8 in. plastic jacket. Digital information is recorded on a magnetically coated surface which is divided into a number of concentric circles called (appropriately) "tracks" and subdivided into "sectors". Each diskette has the capacity of storing 256 K bytes of information.

The disk storage drive unit accepts the diskette through its door opening and rotates the diskette inside its jacket. Access to information on the diskette is made by the magnetic read/write head which lies parallel to the diskette.

Digital information is transferred from the diskette to the computer (or vice versa) by means of two disk controller boards. These interfaces perform synchronization, timing, and direct the read/write head to move to any track or sector of the diskette.

e. Laboratory Information System (LIS)

The Star® System has the capability of communicating with another computer, or output device, via the laboratory information system (LIS). The LIS system is basically a single serial interface board, known as the RS232C. This industry standard device accepts parallel data from the computer and converts it into a serial string of digital information typically sent over a cable. The receiver device may be a teletype, another printer, or smart terminal. The LIS system can be programmed by operator commands to configure itself in meeting the requirements of various existing output devices. Those parameters which may be affected by command are baud rate, start/stop bits, and parity.

The LIS Board occupies one slot position in the card cage. Any report normally fed to the printer may be sent out through the serial port of the LIS system simultaneously.

f. Power Requirements

The Star® system requires only a single electrical connection to be made operational. The power requirements of the system are 117 VAC at 50/60 Hz operation or 220 VAC at 50/60 Hz operation. A 15 A dedicated line is sufficient for the system power requirements.

Five power supplies (+15 V, −15 V, +5 V, +24 V, +24 HV) maintain all DC voltage levels necessary for system operation. A voltage level meter and a rotary switch is provided on the front panel allowing for easy monitoring of all DC voltage levels.

2. Software

All system software is provided on a single diskette. The configuration scheme of the software (as formatted on the diskette) allows the Star® System to have what is known as "virtual memory".

The Star® operating system software may be conceptually divided into four basic types: main, overlays, data files, and results files. Main includes all software for the executive program (scheduling), interrupts (timing and control), I/O routines, format of outputs, and disk handling routines. Main is always present in the system random access memory (RAM). Overlays are executable code programs which are called into the system RAM from the diskette as required. Data and results files are purely data-oriented and are stored on diskette until required.

a. Command Processing

Upon initiation of the start-up state, Star® automatically loads the diskette main contents into the RAM. Once main is loaded, the program alerts the computer to monitor the keyboard for command inputs. Anything typed on the keyboard is echoed directly to the CRT. When the enter button is pushed, the information keyed may be accepted for rejected. If rejected, an error message will be displayed. If accepted, the software will lead the appropriate Overlay from the diskette into the system memory. The CPU will then execute the Overlay program to allow the chosen command to be actualized. There is approximately one overlay for each individual command.

b. System States

Star® System operation is divided into 5 system states:

1. Start-up
2. Ready
3. Operate
4. In-process
5. Shut-down

Although most of the required functions are carried out by system software, manual intervention by means of the keyboard and CRT is required to guide the operation of Star®.

Transitions between states are typically initiated by the operator, but can occur automatically in certain situations. State transitions are primarily a function of the present state, the allowable paths out of the present state, existing system failure (if any), and operator intervention via Keyboard Commands.

The start-up state delivers the system from an Off position to the point where the ready state may be initiated. Start-up state functions are the following:

1. Apply power to analytical and electronics modules once the power on switch is pressed.
2. Routine computer, memory, and disk diagnostics are performed.
3. Main system program is loaded to RAM from disk.
4. System program is initialized.
5. Electro-mechanical devices are initialized.
6. The executive program is started.
7. The checklist is displayed on the CRT.
8. Operator is requested to enter time and date.
9. After entry of time and date, pump module is energized.
10. Computer monitors the keyboard and processes any operator commands.
11. Possible system fault areas are monitored.
12. Automatic error sensor calibration is initiated.

The ready state can be entered twice during overall system operation. It may be initiated at the end of start-up or at the end of in-process. Exits from ready are to operate or shut-down states.

The ready state has two specific functions. The first is increased control of electro-mechanical devices and monitoring the operator keyboard for specific commands.

The operate state is the mode where samples are processed and results reported. It encompasses the time from the initiation of the first aspiration to the end of sampling. The operate state may be entered from either the ready or the in-process states. The operate state is entered when the "operate" button is pressed on the keyboard. Possible exits to other states are to in-process or shut-down states. Primary functions of the operate state are

1. Control the operation of electro-mechanical devices during processing.
2. Aspirate samples, standards, controls, primers and increment the sequence counter.
3. Process any phasing definition inputs from the operator.
4. Monitor the keyboard and process commands.
5. Calculate and display results.
6. Print results or transfer to other output devices.
7. Monitor for system failures.
8. Initiate automatic phasing.

The in-process state is defined as the interval from the last cup sampled to the print-out of results from that last cup. Transition to this state is either by keyboard command or automatically either upon completion of the worklist or upon system failure.

From the in-process state, exit may be to ready, back to operate, or directly to shut-down states. Note that the in-process state performs the same functions listed under operate, except that samples are no longer aspirated, and the sequence number will not be incremented.

One may enter the shut-down state by a keyboard command from any other state, or automatically from any state when a system failure is detected. The shut-down state accomplishes the following:

1. Control electro-mechanical devices.
2. Monitor operator keyboard and process commands.
3. Monitor for system failure.
4. Process operator command for system wash.
5. Shut off power to pump module when entry to shut-down state is due to waste alarm.
6. Restore power to pump module when waste overflow has been corrected.

Figure 6 is a diagram that summarizes the Star® System States and all their possible transitions. Certain conditions that trigger a transition, labeled "Class 2" or "Class 3" in the diagram, refer to certain types of system errors and will be discussed under Section III.2.g, Error Reporting.

c. Control States

Control states are defined in terms of the condition of the hydraulic control valves, the state of the magnets, sampler, pump module and scintillation window, including monitoring of their initial states.

The processing cycle of Star® is 60 sec. It begins with sampling and ends when the bound portion of the reaction leaves the counting device.

Tray rotation occurs when the sample probe has been moved to a wash position after aspiration of 90 μl from the sample cup. Rotation occurs after every sample aspiration in the 60/hr mode, but in the 30 or 15/hr modes, it occurs only after every second or fourth sampling, respectively.

The magnets, detector valve, count gate, and control state timing is accomplished by the triggering of the sensor or light activate switch (LAS) at the exit of the incubation coil.

The energy window of the scintillation detector is controlled by assay definition command and system operator via keyboard. Boundaries are entered for ^{125}I and ^{57}Co automatically and are governed by assay requirements. The operator may establish the high and low energy boundary by entering values within the 15 to 250 KeV range. The

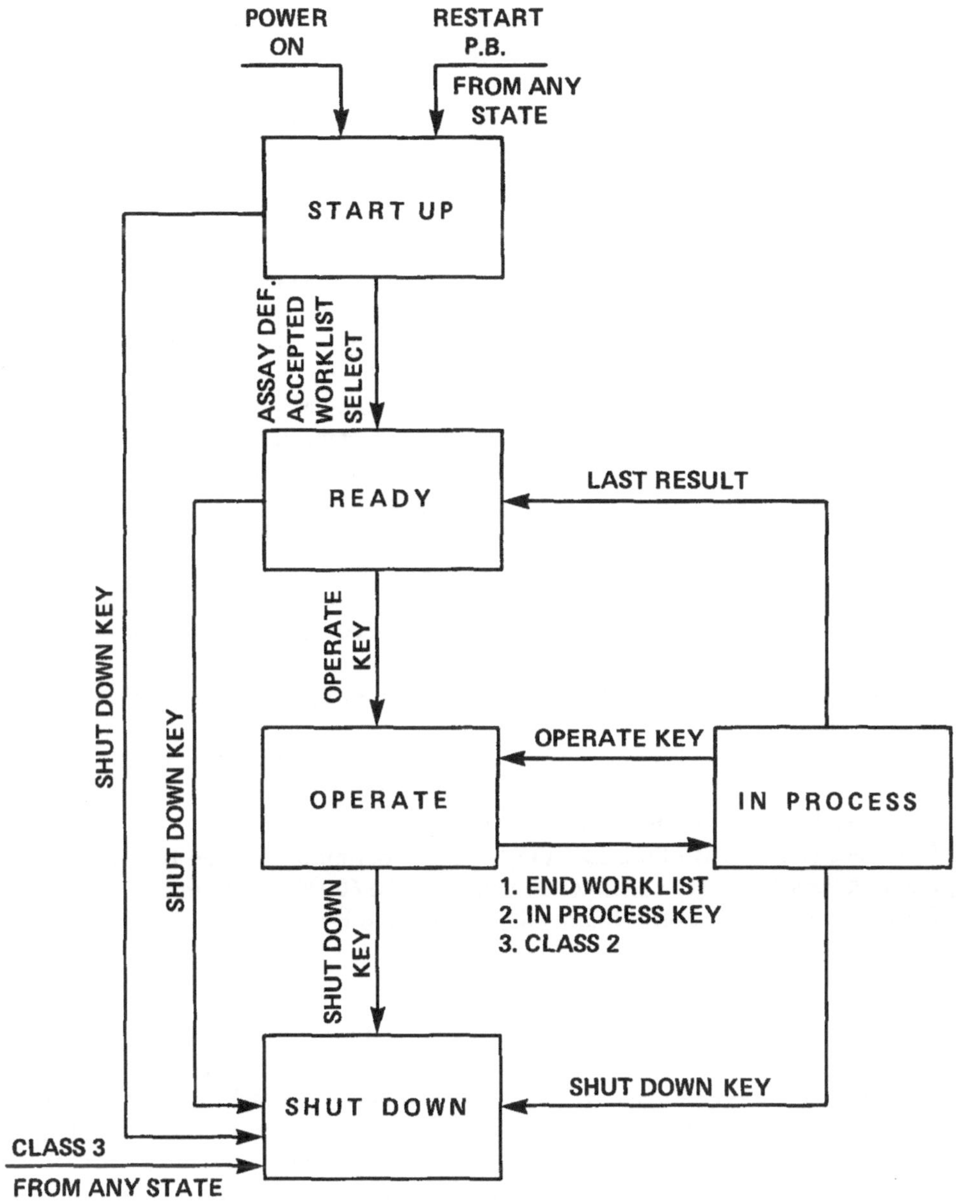

FIGURE 6. Flow diagram of Star® System States and possible transitions.

detector amplifier receives these boundary readings as two DC voltage levels, generated by the digital to analog conversion of two computer outputs.

Within one second after the Main software is loaded from the disk in the start-up state, valves 1 to 8, magnets 1 and 2, and the vibrator are de-energized and the pump signal reset to off.

The sampler is then initialized as follows: the probe is raised, 1 sec later it is moved to wash, the tray rotation control output is reset. Immediately thereafter, the scintillation detector window voltages are initialized so that the low level is 25%, the high level is 75% of full scale.

Sampling is initiated after entry to the operate state and every 60 sec thereafter in the recommended mode. Samples may be aspirated in one of three modes of aspiration, corresponding to one, two, or four aspirations from each cup. These aspirations are all at the analytical rate of 60/hr, but the net rate then becomes 60, 30 or 15 answers/hr dependent upon selection of mode. Standards are always aspirated in duplicate. The selection of replicate modes of operation is primarily a function of the low sensitivity normally previously associated with certain assays.

d. Data Reduction

The Star® System performs data reduction by computing the antigen concentration (dose) of unknown samples. The computation depends upon the scintillation detector event counts for the unknown sample and upon a standard reference curve. The system also measures the coefficients of variation between the computed standard curves and the curve computed from measurements made on samples of known dose. All data reduction is performed in the operate and in-process system states.

Success in data reduction thus depends upon event counts, dead time counts, standard dose values, sample type code, and processing mode inputs.

The computer reads the 16-bit scintillation counter every 40 msec. This reading is actually the event counts and they are accumulated over a 45 sec interval within each 60 sec aliquot cycle. Another scintillation 8-bit counter is also ready every 40 msec by the computer. This reading is known as the dead time counts and they are accumulated over the same 45 sec processing cycle.

Up to nine standard dose values are obtained from operator inputs via keyboard for assay definition.

The sample type code is obtained from the worklist and is entered by the operator during start-up state. This code indicates that the sample being processed is either a standard, an unknown, a control, or a primer.

The processing rate is obtained from the assay definition entered by the operator also during the start-up state. As mentioned, it may be 60, 30, or 15 samples/hr.

The dose or concentration is computed per Equation 1:

$$\text{Dose} = \text{Inverse Log of } (Y - A0)\ (1/A1) \qquad (1)$$

Where Y = the logit of B/B0 or ln ([B/B0] / [1- B/B0]), B = Net Counts, B0 = the value of B for the Standard having Zero dose, A0 = the Y axis intercept, A1 = Slope.

The standard curve coefficient is calculated as follows: for each standard, excluding the zero dose standard, the log of the standard dose, Xi, and the logit of B/B0, Yi, are determined. Once the complete set of N values for Yi are determined for each Xi, the standard curve coefficients are determined from a linear regression. Equation 2:

$$\text{Slope (A1)} = \frac{N \Sigma\, XiYi - (\Sigma\, Xi)\,(\Sigma\, Yi)}{N \Sigma\, Xi^2 - (\Sigma\, Xi)^2} \qquad (2)$$

Equation 3:

$$\text{Y intercept (A0)} = Y - A1\,(X) \qquad (3)$$

Where X and Y are the means of Xi and Yi values, respectively.

Once the coefficients of A1 and A0 have been determined, the dose can be computed for each standard using the dose equation above.

The net count (B) determination is different for each processing rate; at 60/hr, there is only one 45 sec count interval, therefore, net count (B) equals the current net count.

At the 30/hr rate, however, there are two 45 sec count intervals, therefore, by Equation 4:

$$B = 0.5\,(B1 + B2) \tag{4}$$

Where B1 is the current net count when the first count interval has expired and B2 is the current net count when the second count interval has expired.

At the 15/hr processing rate, there are four 45 sec count intervals, therefore, by Equation 5:

$$B = 0.5\,(B3 + B4) \tag{5}$$

Where B3 is the current net count when the second interval has expired and B4 is the current net count when the fourth count interval has expired.

The current net count (CURNC) is determined for an interval of 45 sec of event counts (EC) and dead time (DT). Every 40 msec the CURNC is calculated by Equation 6:

$$\text{CURNC} = K\,\frac{\text{ECSUM}}{1 - \text{DTSUM}/\text{CI}} - \text{NSC} \tag{6}$$

Where K is a constant of 0.1 during background measurement and 1.0 at other times, where ECSUM is accumulated net event counts, DTSUM is accumulated dead time counts, NSC is the nonspecific count or background, and CI is the elapsed count interval.

Star® performs certain other computations. The correlation coefficient is a measure of the degree to which two variables relate to each other (i.e., specified dose and computed dose). The correlation coefficient, or r, is determined by Equation 7:

$$r = \frac{N\,\Sigma\,(SD)\,(CD) - (\Sigma\,SD)\,(\Sigma\,CD)}{\sqrt{N\,\Sigma\,(SD)^2 - (\Sigma\,SD)^2}\;\sqrt{N\,\Sigma\,(SD)^2 - (\Sigma\,CD)^2}} \tag{7}$$

Where r = Correlation coefficient, N = Number of individual values, SD = Specified dose (standard), CD = computed dose (standard).

A percent difference is also determined for unknowns and standard samples according to the following Equation 8:

$$\text{Percent Difference} = \frac{(B1 - B2)}{B1} \times 100 \tag{8}$$

Where B1 and B2 are NET COUNT outputs.

All data reduction outputs are formatted for printout, CRT display and transmission via serial channel (LIS) to other devices. Table 1 charts the availability of the particular data outputs to various system destinations.

e. System Monitoring

System Monitoring is an inherent requirement of automated instruments whereby the operator is kept constantly informed of the system status during the course of operation. Monitoring of the Star® System alerts the operator to hydraulic malfunction, which is subsequently reported by the CRT and/or on the printout. Also, the

Table 1
DESTINATION OF DATA REDUCTION OUTPUTS

Type of output	CRT	Destination Printer	Serial
Dose	X	X	X
Computer dose (standard)	X	X	X
B/B0	X	X	X
Net count	X	X	X
Average net count	X	X	X
Current net count	X	—	—
Std. curve coefficients (Slope & intercept)	—	X	X
Percent difference	X	X	X
Correlation coefficient	—	X	X
Background count	—	X	X

system will designate which of the failure modes may have occurred and act accordingly to either continue or abort operation. System monitoring also involves the control of timing relationships affecting reagents in the analytical stream (phasing).

As noted before, at the heart of the Star® Monitoring System are the four light activated sensors. Three of the sensors are utilized to detect hydraulic malfunction and monitor the phased addition of reagents. One of these sensors is dedicated to the sample line, a second controls the label line, and the third monitors the solid-phase flow. The fourth sensor controls the incubation outflow and initializes the timing of the magnets, vibrator, detector valve, and the count gate by the detection of the solid-phase particles in the reaction stream.

Error detection is accomplished by reading each of the three sensors at their respective absorbence levels. The sample sensor monitors sample, air, and buffer. The label sensor monitors label, air, and buffer. The solid phase sensor monitors that substance, air, and buffer. Preset absorbence levels of all reagent constituents passing through the detectors are designated as a percentage of full scale, where full scale is 100% absorbence. Table 2 lists these levels of absorbence at which the sensors are set. The absorption of each stream being sensed is converted to a voltage signal. This voltage is inversely proportional to the log of the absorption. At this point, the voltage is converted to a one byte digital value via analog to digital (A/D) conversion for input to the computer. Each of these digital inputs is sampled by the computer once during every 200 msec interval.

To prevent bubbles from confusing the level detection algorithm, groups of readings are taken that correspond to a two second bubble duration. The highest value within each group is then taken as the reading for that group. This is called the debubbled value. Thereafter, each sensor reading is compared to a threshold value which is utilized as a reference point and is referred to as the detect level. As each constituent passes by the sensor, the stream is read and a comparison of the detect level and the debubbled value is made to verify that the sensor reading is at the correct (expected) absorbence level for that particular reagent constituent. The sensors through the algorithms associated with them identify the following flow error conditions:

1. No sample
2. Short sample
3. Long sample
4. Clogging in the line

Table 2
ERROR DETECTION ABSORBANCE LEVELS

Reagent constituent	Absorbance level
Sample	80 ± 10%
Label	80 + 10%
Solid phase	0 + 5%
Buffer	60 ± 10%
Water	80 ± 10%
Air	35 ± 5%

5. Short label
6. Long label
7. Short solid phase
8. Long solid phase
9. Depletion of buffer

The Incubation Outflow Sensor is used primarily to time all events following passage of the stream out of the incubation coil. The debubbled value for this sensor is determined identically as above, with the exception that this sensor is keyed to the detecting of a transition only when solid phase is present. The presence of a false transition is controlled by the algorithm used and does not trigger the sensor timing dependent devices.

The Star® performs automatic phasing by monitoring the same flow sensors primarily dedicated to error detection. To provide ideal reaction conditions and precision, an aliquot of sample is allowed to enter the reaction stream. The label valve is then activated which allows the labeled antigen to fall on top of the sample stream. After another interval of time, the solid phase valve is activated and solid phase particles are released on top of the sample and label mixture. This scheme supplies a constant amount of sample, label, and solid phase to the reaction stream.

Phasing is referenced directly to the sampling duration of 34 sec. Label and solid phase sampling times during the 60 sec cycle are 26 and 18 sec, respectively. The system is defined as accurately phased when the label is delayed by 4 sec and the solid phase is delayed by 8 sec, both with respect to the sample timing.

The pyramid effect thus achieved is created by the computer through accurate timing of the label and solid phase valves. This is achieved by energizing the label and solid phase valves to advance or retard the flow with respect to the sample as reference. The computer energizes the valves according to the measurements read from the sensors and the following Equations 9 and 10:

$$\text{Label Delay} = Ts + T1 - (T_L + T_2 - 4) \quad (9)$$

$$\text{Solid Phase Delay} = Ts + T1 + T3 - (Tsp + T4 - 8) \quad (10)$$

Where Ts is the time from start of sampling to the sample sensor time. This time interval is determined by the computer.

T1 is the time from sample sensor to the injection point 1(IP1) — a constant. TL is the timing of label from label valve to the label sensor. This value is determined by the computer. T2 is the time from label sensor to the injection point 1 (a constant).

Tsp is the time interval from solid phase valve to solid phase sensor. This is determined by the computer. T4 is the time from solid phase sensor to the injection point 2 (IP2) — a constant. T3 is the time interval from the injection point 1 to injection point 2. This value is to be determined by the system operator and entered via keyboard.

Automatic phasing commences during the operate state where three aspirations are taken from the primer cup. The first aspiration primes both solid phase and label valves to ensure a sufficient volume in the valves' stream. The second aspiration accomplishes the phasing by determining the values of phasing parameters TL, TS, and Tsp times as detected by the sensors.

f. Data Monitoring

Star® sets normal limits on standard and unknown sample values and subsequently flags all results determined to be beyond those limits. The system responds according to the severity of the error recorded. Error messages are indicated on the CRT and/or printout and will identify the data error by type.

The following parameters will be checked on standards by the computer:

1. B1/B0 or the ratio of the first count and the average net count for the zero standard
2. B2/B0 or the ratio of the second count and the average net count for the zero standard
3. The percent difference between net count output of standards
4. The percent difference between the specified dose and the computed dose
5. The correlation coefficient (r) between the specified dose and computed dose
6. The slope (A1)
7. The Y intercept (A0)

The following parameters are checked on unknown samples:

1. Dose — the computed dose is checked with respect to the normal range as input by the operator. The dose is also compared to the assay range.
2. The net count ouptut is compared with the zero standard.
3. The percent difference between net count output of unknowns at 30 and 15/hr.

g. Error Reporting

Star® has the ability to indicate system errors by display, printout or audible alarm. System failures are defined by class as follows:

Class 1 — A sample specific problem will generate this alarm. The operator is notified of the alarm by both audible and visual indications and the results will be flagged. No transition out of its state will occur. The operator makes decisions concerning this failure and the monitoring of the problem. The system will continue to execute the unaffected functions. The audible and CRT visual alarm indicators will remain ON until the operator acknowledges by keyboard controls.

Class 2 — Under this failure class some or all samples in the system are doubtful (e.g., depletion of reagents and/or clog). The system will indicate this failure by both visual and audible means. All results determined after time of failure will be flagged and the system will be automatically transferred to the in-process state. Operator intervention is required to correct this failure and the visual alarm will remain ON until the problem is corrected. The audible alarm will remain until it is shut off by operator acknowledgment.

Class 3 — This failure will result in the total shutdown of the system and will automatically shut-off power to the pump module only. The system will indicate this by visual and audible means. Operation will only be continued after the corrective measures have been taken. To continue operation, a system start-up or shut down state must be initiated. The Class 3 failure will occur under such conditions as full waste bottle or leakage in the hydraulics.

Not Classed — There are a number of errors which are not classed because they do not fall in any of the above categories.

All errors produced will be classified as follows:

- System errors
  - Flow errors
    - LAS
    - Short sample
    - Sample
    - Solid phase
    - Label
  - Disk errors
  - Waste
  - Mixer
  - No clock
- Data check errors
- Entry errors
- Start-up diagnostics errors

Flow errors — Flow errors are primarily the result of depletion, delay or no flow of the various constituents and/or clogging in each respective flow sensor stream. Flow errors are classed as either 1 or 2 depending upon the number of occurrences.

Disk errors — Disk errors can occur at any time during operation. They indicate that one of eleven different failure modes have been indicated in the flexible disk system. These failure occurrences will be displayed on the CRT.

Waste — A waste error message will be indicated on the CRT and Printout when the liquid waste level has reached specified limits. The class error is 1, 2, or 3 dependent upon the number of timed occurrences the error is detected.

Mixer — The mixer motor alarm is indicated when attempting to operate with the mixer motor off.

No clock — No clock is a system timing error. This error will occur when no interrupt has occurred from I/O Board 1. The CPU may become disabled; the system remains in its present state.

Data check errors — Refer to the data monitoring paragraph for data checks performed on standards. Based on the ratio of the first and second counts to the average net count for the zero standard, specific count failures are determined on any combination of the counts. The system will give a Class 1 or 2 error dependent upon the gravity of the count failure. The # symbol and the count failure(s) message is outputted when the standard curve is invalid and needs to be repeated. The percent difference, slope, intercept and correlation coefficient will produce a Class 2 error and output a STD curve failure message when preset limits have been exceeded. For unknowns refer to data checks performed. All unknown data errors will produce a Class 1 error. For dose determination a dose value which has exceeded the normal range will be identified by the asterisk * symbol. A dose value which has exceeded the assay range will be identified by a less than < or greater than > symbol followed by the lower or upper limit value of the assay exceeded respectively. The net count and percent difference will be identified by the asterisk symbol, when they exceed the specified limit. The background count will be flagged by the # symbol when exceeding a background count of 100.

Entry errors — Entry errors will only be indicated on the CRT and will indicate the specific error caused by entering an improper command code, improper statement, format, illegal state transition and nonacceptance of parameters.

Start-up diagnostic errors — Start-up diagnostic errors (if any) will be indicated on the CRT and Printout when power on or reset is initiated. The system performs automatic diagnostics (CPU, memory, disk) on start-up. If an error is indicated, a message will give the specific error location. If no errors occur, messages will be followed by OK. If a CPU error is indicated, then a halt has occurred and no CRT or printer message will be outputted. Table 3 is a complete list of error messages and their mode of display.

1. The circuit breaker supplies and removes main power to all units.
2. The emergency button removes the main power from all units and trips the circuit breaker.
3. The on-off power button controls power to the system (the detector PMT is always on) and is used for normal on-off system.
4. The reset push button is a computer reset button which will deliver the system to the start-up sequence and initialize all program material.
5. The mixer on-off toggle switch (located near the mixer on analytical unit) will control power to the mixer motor.
6. The volt meter rotary switch is used to check the power supplies voltage levels.

IV. OPERATION

A. Operation Flow Chart

The operator interface defines the system in relation to operator control and interaction. Refer to Figure 7 for the operation flowchart of Star®.

B. Control Keys

The control keys are basically divided into three categories: CRT edit keys, Star® command keys, and Star® control buttons. The first two are located on the keyboard, the latter are primarily located on the electronics module front panel. The CRT edit keys allow the editing of characters present on the CRT. Their basis functions are: clearing an entry, positioning the cursor, deleting and inserting characters and scrolling of lines up or down. The Star® command keys perform the following functions:

1. The alarm reset key shuts off the audible alarm and the associated visual error.
2. The printer hold key delays printing of results.
3. The resume print key removes the hold applied by the printer hold command.
4. The print screen key will print the current CRT display onto printer.
5. The paper advance key advances the print paper several lines.
6. The pump control key controls the pump module (On-Off).
7. The enter key is used in conjunction with (and after) all command codes and will allow commands entered through the keyboard to be processed.
8. The operate key is a command to the computer to deliver the system to the operate state, initiate sampling and commence result determination.
9. The in process key is a command to stop sampling. This command will deliver the system to the in-process state.

Table 3
STAR® ERROR MESSAGES

Error reported	Class error	Symbols on printout	Symbols CRT	Audible alarm
A. System errors				
1. Flow errors				
a. LAS	1 or 2	LAS	LAS	Class 2 only
b. Short sample	1 or 2	S. Sample	S. sample	Class 2 only
c. Sample	1 or 2	Sample	Sample	Class 2 only
d. Solid phase	1 or 2	S.P.	S.P.	Class 2 only
e. Label	1 or 2	Label	Label	Class 2 only
2. Disk errors	Not classed	—	Disk error (11)	Yes
3. Waste	1, 2, or 3	Waste	Waste	Yes
4. Mixer	Not classed	—	Mixer	Yes
5. No clock	Not classed	—	No clock	—
B. Data check errors				
1. Standards				
a. Specific count failures	1 or 2	# Symbol count failures or Std. Curve failure	# Symbol count failure(s) or Std. curve failure	Class 2 only
b. % diff.	2	Std. curve failure	Std. curve failure	Yes
c. Slope	2	Std. curve failure	—	Yes
d. Intercept	2	Std. curve failure	—	Yes
2. Unknowns				
a. Dose				
1. Normal range	1	*	*	—
2. Assay range	1	≶	≶	—
b. Net count	1	*	*	—
c. % Diff.	1	*	*	—
3. Background	2	#	#	Yes
C. Entry errors	Not classed	—	Specific entry errors	Yes
D. Start-up Diagnostics Errors				
1. Disk search	Not classed	Disk error	Disk error	Yes
2. Disk read/write		Disk error	Disk error	Yes
3. Memory		Memory error	Memory error	Yes
4. CPU		None	None	—

10. The shut-down key is a command to deliver the system to the shut-down state. A system wash procedure is initiated. The Star® System Control buttons and switches and their functions are as follows:

C. System Commands

Operation of the Star® system is accomplished by using alphanumeric keys and/or control keys to form system commands which the computer can accept. System commands (in most cases) are formed by keying a two letter command code, followed by pressing the enter key. All command and statement codes will be displayed as entered on the entry line of the CRT (bottom-most line).

When the computer accepts the system command, the system will perform one or

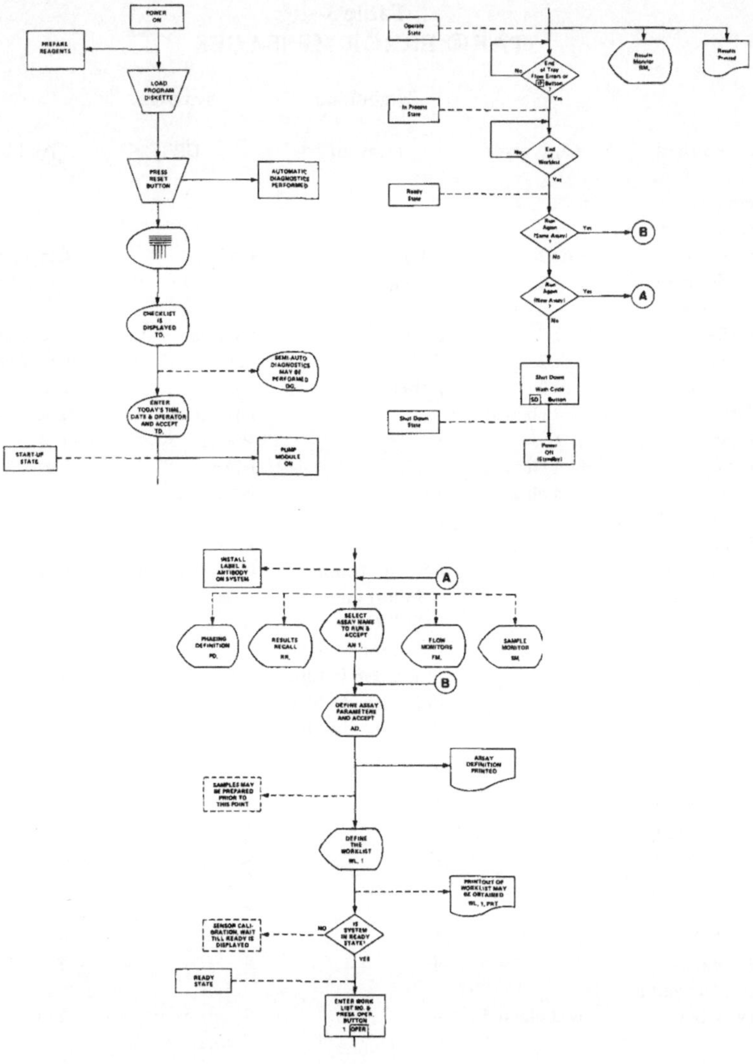

FIGURE 7. Operation flowchart for the Star® System.

more given functions. In most cases the primary command entered will also invoke the corresponding CRT display. Some of the system commands are as follows:

Time and date — The time and date command makes possible the programming of the time, date and operator name onto the printer output associated with data acquisition. The changes are made while viewing its corresponding display.

Checklist — The checklist is a table of preliminary checks which the operator may find useful prior to the system operation.

Assay names — The assay name command allows the operator to select the assay which is to be run from a list of predefined Technicon assay names initially preprogrammed. A means of adding or deleting from the list is possible for accommodating customer defined assays.

Assay definition — The assay definition command and its derivatives allow the user to define, program, and store assay specific data and options on the CRT. This command will define the name of assay, mode of operation, units of measurement, normal ranges, standard values, isotope type and sensitivity, as well as the lot number which appears on the results printout.

Worklist — The Star® System provides the ability to generate a worklist prior to running samples. By use of the worklist commands provision is made to enter standards, unknowns, primer, and control values for each corresponding sample position. The command will invoke the worklist display and allow editing. The results monitor and printout will identify the results with the worklist/sequence number, sample type and optional alphanumeric field. The worklist can also be used to stop the sampler at the end of a run of given length.

Results monitor — The results monitor offers the ability to view results on the CRT as they are being generated. The results monitor is automatically displayed when entering the operate state via the operate key. All pertinent information for each sample as identified by the worklist is displayed along with data output. The display differs with the type of sample being processed. The value for the current net count or running count is displayed on a real-time basis on the upper right of the screen.

Phasing definition — A real-time graphic representation of the phasing is represented on the CRT by this command.

Diagnostics — The diagnostics command allows semiautomatic control of electromechanical devices (such as valves, probe, sampler rotation, pump, magnets, alarm) for trouble-shooting purposes. This command structure will allow either single device control (on-off) or cycle control which activates all devices sequentially for a specified time. The corresponding display allows visual simulation of the devices being activated.

Results recall — The Star® System has the ability to store results on the disk for retrieval at the operator's convenience. These files may be retrieved by worklist and date both visually and on hard copy by the results recall commands. A directory of what is available in the results files may also be obtained.

V. PERFORMANCE

Star® performance characteristics are summarized in Table 4. Further results will be available when formal evaluations are concluded.

Table 4
STAR® PERFORMANCE SPECIFICATIONS

Analysis rate:
: 60 aspirations per hour; standards must be in duplicate, unknowns and controls can be in single aspirations, duplicate or quadruplicate; if unknowns are run in quadruplicate, standards are in quadruplicate as well

Samples/Tray:
: 40 sample cup capability for standards, controls and/or unknowns

Sample container:
: Any standard Technicon sample cup can be used

Sample to wash ratio:
: 1.3:1 (34 sec sample, 26 sec wash)

Sample size used:
: 90 μl per aspiration

Throughtime:
: 15 min from aspiration to result

Incubation temperature:
: 37°C ± 0.5°C

Antibody volume:
: 47 μl/aspiration

Radiolabel volume:
: 68 μl/aspiration

Table 4 (Continued)
STAR® PERFORMANCE SPECIFICATIONS

Required environmental conditions:
  Ambient temperature: 18°C—32°C
  Relative humidity: up to 70% (noncondensing)
Power requirements:
  Line voltage to the system should be 105—125 VAC at 60 Hz
Weight:
  Sampler: 12 lb
  Analytical module: 225 lb
  Electronics module: 205 lb
Space requirements:
  Height: 23 in.
  Width: 72 in.
  Depth: 23 in.
Gamma detector:
  2 in. × 2 in. well crystal with plastic helical coil and 1 in. lead shielding front and sides with ¼ in. shielding in rear
Detector efficiency:
  76% DPM = CPM
Gamma detector counting time:
  45 sec/aspiration
Carryover:
  Less than 1% between specimens
Coefficient of variation:
  Less than 5% in normal range
Data output:
  Display on Cathode Ray Tube for samples in progress and one or two copy pressure sensitive paper printout with tray and cup sequence number, patient name and or number (up to 14 characters), counts per sample from the bound portion of the reaction, average of duplicate CPM (if run in duplicate), percent difference of duplicates from the mean of the CPM, and concentration of the specimen
Computer interface:
  RS232C included, output is serial ASCII
Waste disposal:
  Up to 8 hr capacity in a waste container with alarm 40 min prior to full
Alarms:
  Waste bottle full, short sample, reagent line empty, duplicates out of range, standard curve checks for slope and linearity, out of range for unknowns, high background
Assays available at initial shipments;
  T_4
  T_3 Uptake
  T_3 (RIA)
  Digoxin
  Cortisol
  Human placental lactogen (HPL)

REFERENCES

1. **Erlanger, B. F., Borek, G., Beiser, S. M., and Lieberman, S.,** Steroid-protein conjugates. II. Preparation and characterization of conjugates of bovine serum albumin with progesterone, deoxycorticosterone and estrone, *J. Biol. Chem.*, 234, 1090, 1959.
2. **Snyder, L., Levine, L., Stoy, R., and Conetta, A.,** Automated chemical analysis: update on continuous flow approach, *Anal. Chem.*, 48, 942A, 1976.

Chapter 5

GAMMAFLO^{TM}: AUTOMATED ASSAY SYSTEM

D. R. K. Murty, S. W. Hitch, and G. Brooker

TABLE OF CONTENTS

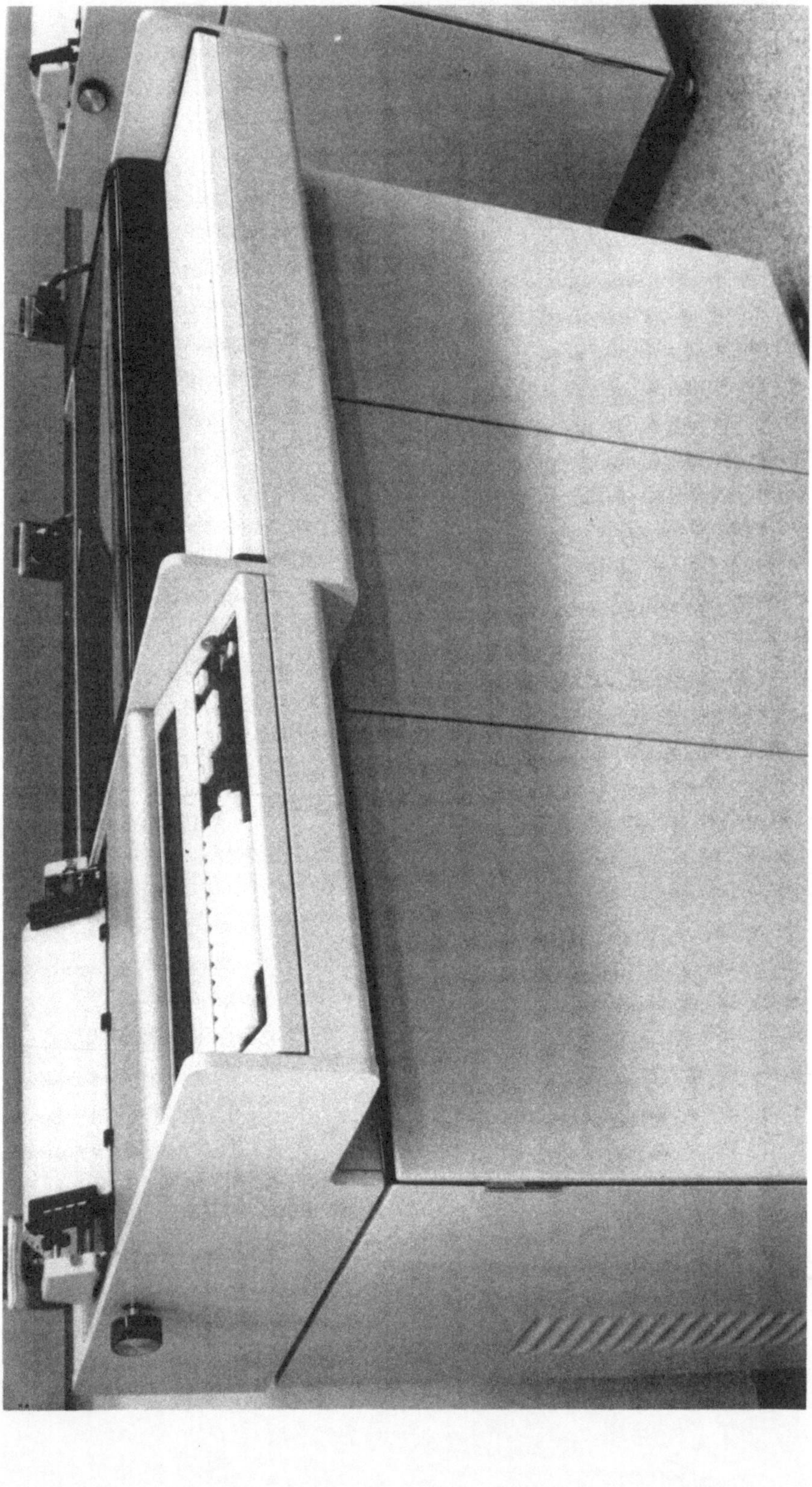

Figure 1. The Gammaflo ® Automated Assay System.

I. INTRODUCTION

The widespread utilization of the radioimmunoassay technique as a sensitive, specific, and accurate method for the measurement of a wide variety of compounds of biochemical and clinical interest has stimulated interest in the development of automated technology. Although systems that could mechanize portions of the tedious, exacting manual procedure were emerging, no technique or equipment provided complete automation.

The basic concept of the Gammaflo™ Automated Assay System originated at the University of Virginia in Charlottesville. The design and engineering of a reliable, completely self-contained, computer-controlled instrument and the development of reliable reagents for clinically important assays has been accomplished at the Squibb Institute for Medical Research, Princeton, New Jersey. It is the purpose of this chapter to describe the Gammaflo™ System.

II. THE GAMMAFLO™ CONCEPT

The Gammaflo™ System is a hybrid combination of continuous flow methodology and discrete sample analysis. The instrument consists of a peristaltic pump, a sample pickup device, a sodium iodide gamma-radiation detector, a column to separate antibody-bound from free radioactivity, associated miniature solenoid valves to control solution flow, and a microcomputer to calculate results and to control the instrument. The system accurately combines known volumes of sample, radioligand, and specific antiserum, incubates the reaction mixture for a precise length of time, separates and counts the antibody-bound labeled ligand, and makes all necessary calculations.

The initial sampling and mixing of reagents is accomplished by a continuous flow design that has proven to be accurate, simple, and flexible in many colorimetric procedures. The Gammaflo™ concept, however, departs significantly from conventional flow systems in that reactants are discretely introduced with each sample thus conserving valuable reagents. The radioactive sample (reaction mixture) itself triggers the valve timing sequence that routes the reaction mixture through the separating and counting portions of the system. Analysis time is decreased because the reaction mixture, once detected, is sped on its way by means of a high-flow buffer mechanism. A resin-containing column is used to separate free from antibody-bound labeled ligand. After each separation, the column is automatically washed and prepared to receive the next sample mixture. Counting accuracy is maximized by a stopped-flow technique in which the antibody-bound radioligand is held stationary in the detection well and quantitated. The sample is then rapidly evacuated from the detection system.

Figure 2 is a schematic diagram of fluid pathways in the Gammaflo™ System. The pipetting of radioligand, antiserum and sample is accomplished by the sample/reagent pickup arm. The computer directs the arm to move from its resting state (during which all three lines draw buffer) to its sampling position for the pipetting of sample, antiserum, and radioligand. This action is precisely timed by the computer, thereby yielding uniform volumes of reagents and sample.

The sample, antiserum, and radioligand move through the pump to a mixing tee where they are combined and mixed with incubation buffer and air bubbles. The air bubbles are added to reduce laminar flow which could cause mixing of sequential samples. The reaction mixture then proceeds through the incubation coil. Because the peristaltic pump provides a constant flow rate, the incubation time from sample to sample is uniform.

After passing through the incubation coil, the sample-reagent mixture moves

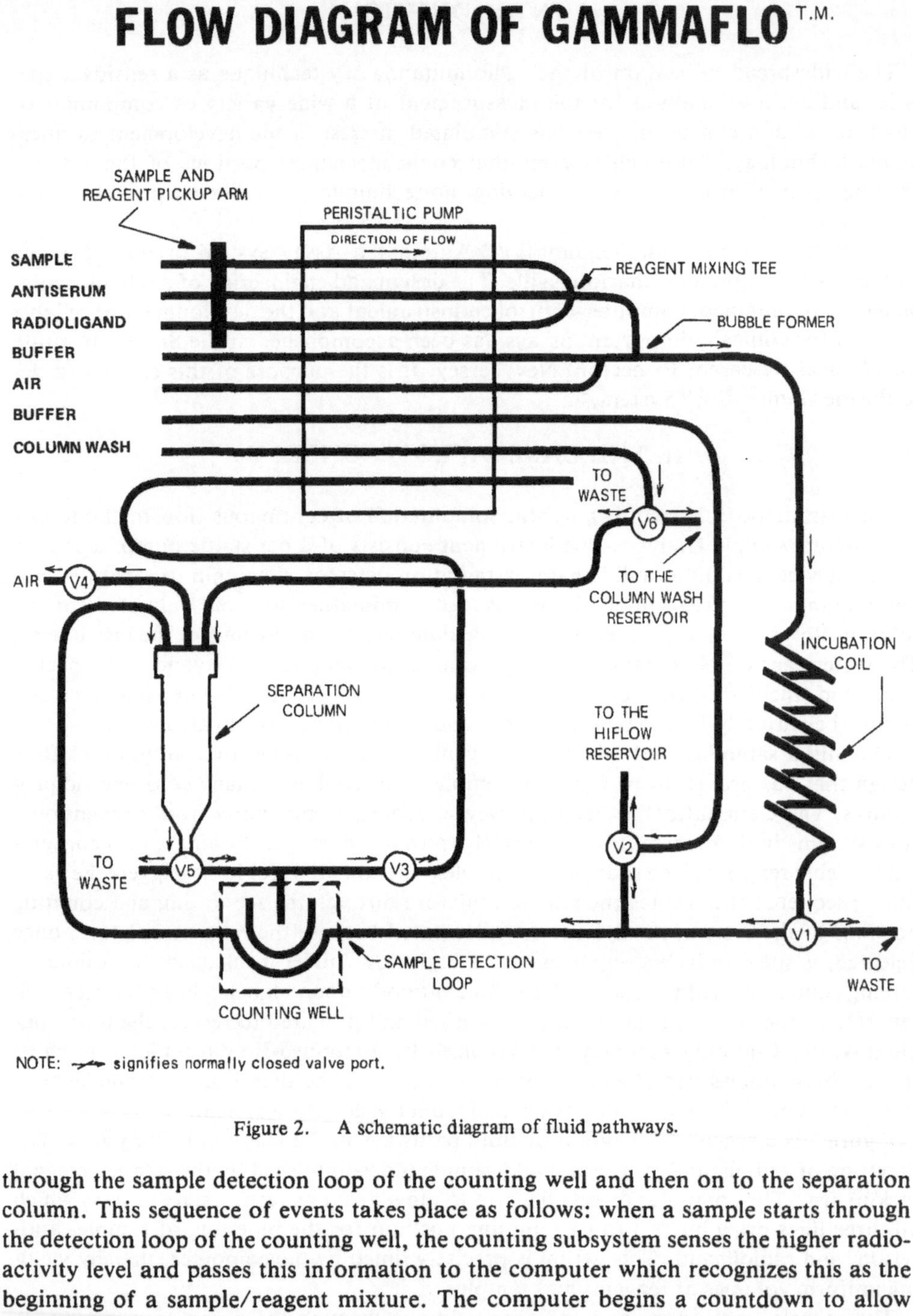

Figure 2. A schematic diagram of fluid pathways.

through the sample detection loop of the counting well and then on to the separation column. This sequence of events takes place as follows: when a sample starts through the detection loop of the counting well, the counting subsystem senses the higher radioactivity level and passes this information to the computer which recognizes this as the beginning of a sample/reagent mixture. The computer begins a countdown to allow time for all of the mixture to pass valve 1. Once this has taken place, valve 1 is switched so that the intersample buffer is directed to waste and valve 4 is closed so that the column becomes pressurized. Simultaneously, valve 2 is opened so that the high-flow buffer pushes the sample mixture through the column. Valve 5 switches so that the antibody-bound material passing through the column is directed to the counting well. After a period of time controlled by the computer, the column is washed in preparation for the next reaction mixture. Valve 5 switches to waste and valve 6 opens pushing column wash fluid through the column. Valve 6 then closes and valve 2 reopens so that a burst of high-flow buffer can reequilibrate the resin to its initial state. While the column is being washed, the amount of antibody-bound radioligand in the counting well is quantitated. After the sample has been counted, valve 3 opens and the sample is withdrawn by vacuum from the counting well. Valves 1, 2, and 4 return to their original positions and the analytical portion of the instrument is ready for the next reaction mixture.

The count is then processed by the computer as either a standard or an unknown. If the sample is a standard, the count is stored until all the standards specified in the assay protocol have been acquired. These counts are then used to develop a standard curve, which is plotted on the instrument's printer/plotter. If the sample is an unknown, the quantity of unlabeled antigen in the sample being analyzed is determined by comparison to the standard curve and the result is printed.

III. THE GAMMAFLO™ SYSTEM — DESIGN GOALS

The Gammaflo™ System is composed of the instrument and the reagents. A number of design goals were established at the outset of instrument development. Chief among these were reliability, versatility, and supportability. Achievement of these design goals was greatly enhanced by utilizing a powerful, highly integrated computer.

The need for reliability in the clinical laboratory instrument is well understood, yet reports of excessive down time among clinical instruments are common. The Gammaflo™ concept lends itself to a reliable implementation because there are a limited number of moving parts. All components utilized in the Gammaflo™ instrument have been extensively tested and proven reliable. It was determined early in our development efforts that the most vulnerable components in our prototype instruments were the valves. After an extensive review of alternative valve designs, we discovered a design which is routinely capable of over one million operations without failure. The other major electromechanical components utilized in the Gammaflo™ instrument are the peristaltic pump, the printer/plotter, and the disk drive. In each case, selection was made on the basis of reliability rather than cost or capacity. The remaining electromechanical components are the pipette mechanism and the sample tray. These components are driven by highly reliable stepping motors under computer control. Both the pipette mechanism and the sample tray provide feedback signals to the computer to verify that the operation instructed by the computer has been successfully performed.

Special care has also been taken in the development of the instrument's electronic circuitry to enhance its dependability. Potentially troublesome connectors were held to a minimum by consolidation of as many circuits as possible on an individual printed circuit board. Potentially conflicting electrical signals are carefully isolated from one another. A built-in isolation transformer reduces the instrument's vulnerability to aberrant conditions on the user's power lines. The backbone of the system is an established computer of proven reliability.

Since the Gammaflo™ instrument is computer-controlled, software plays a fundamental role in instrument reliability. The Gammaflo™ software is written in structured code and is being developed to run under a real-time operating system of proven reliability.

The Gammaflo™ System has been designed to provide flexibility and adaptability both in the present and in the future. The Gammaflo™ instrument provides the user with considerable control over assay protocol. The user has the ability to change virtually every parameter in the assay sequence: replicate number, sample and reagent volume, sample interval, and valve timings. Additionally, the user has the ability to modify the format of the printed report, dividing the sample identification field into as many as four identification columns. For instance, the user could thereby include patient name, identification number, room number and admission date on the printed report.

It was our objective to develop a comprehensive set of data algorithms for use with Gammaflo^{TM}. The following curve fitting techniques have been implemented: logit, logit log, endpoint adjusted logit log, weighted logit log, third order polynomial, and spline under tension. Large, easily readable curves are plotted on the instrument's printer/plotter allowing the user to visually assess the quality of their fit. The appropriate statistical information for this purpose is also computed.

An outstanding feature of the Gammaflo^{TM} concept is that it uses conventional reagents: no special coding processes or special materials are utilized. As a result, the Gammaflo instrument lends itself to the use of reagent systems developed by the users. All of the diagnostic features designed to expedite reagent development by Squibb scientists have been left in the instrument to facilitate assay development by the user community.

The Gammaflo^{TM} instrument is designed to avoid obsolescence. Virtually the entire control logic is stored on a floppy disk. Therefore, insertion of a new disk instantly updates the system.

IV. THE GAMMAFLO^{TM} SYSTEM — IMPLEMENTATION

The upper portion of the Gammaflo^{TM} instrument is divided into two sections. The left-hand portion, which is shown in Figure 3, is the operator's console through which the operator communicates with the instrument. It contains a standard keyboard, operation keys, a display, various status lights, and the printer/plotter. The right-hand section of the upper portion of the instrument, which is portrayed in Figure 4, is the chemistry section. It contains all the components employed in implementing the basic Gammaflo^{TM} concept. With the exception of a short work surface at the front of the instrument, this entire section is covered by a transparent cover. The lower portion of the instrument is divided into three sections (each covered by its own door). The area below the operator's console contains paper for the printer/plotter, the dual-drive floppy disk system, and the power distribution panel. The center section contains the microcomputer subsystem and portions of the nuclear subsystem. The right-hand section contains the bulk reagent and waste tanks.

As previously noted, a sampling system, peristaltic pump, incubation coils, valves, separation column, and counting system are utilized in implementing the Gammaflo^{TM} concept. During the next several paragraphs these components will be described in more detail.

The sampling system consists of three components: the sample positioner, the sampler/reagent pipet mechanism, and the reagent block. The sample positioner aligns the sample for pickup by the pipette mechanism. The visible portion of the sample positioner consists of 35 racks, each holding five cuvettes, thus providing a capacity for 175 samples. Because Gammaflo^{TM} can be programmed to sample as many as 25 times from the same cuvette, the positioner has a capacity for 175 *different* samples. Therefore, if duplicates are run, the effective number of samples would be 350 in comparison to an instrument without this capability. The cuvettes utilized in the Gammaflo^{TM} instrument are of 1.5 and 0.5 mℓ capacities and are available from a number of independent distributors. The sample positioner is a compact, well-proven design. The mechanism is driven by a stepping motor, and optical sensors are utilized to provide

Figure 3. The operator's console including the keyboard, operations keys, display, and printer/plotter. A typical printout is shown.

sample positioning. The sample positioner is under computer control; special feedback circuits virtually eliminate the possibility of sample sequence errors.

The sample/reagent pipet mechanism consists of three probes that move between the forward or pipetting position (see Figure 5) and the resting or buffer position (see Figure 6). In the latter position the probes are rinsed in order to avoid intersample contamination. Its unique motion allows the use of narrow mouth reagent bottles, thereby reducing evaporation and dust contamination of the reagents. Like the sample

Figure 4. The chemistry section including the sample positioner, the sample-reagent pickup mechanism, the reagent block, the peristaltic pump, the incubation coils, the gamma counter, the separation column, the valves, and the vacuum chamber.

Figure 5. The sample-reagent pickup mechanism in the forward or pipetting position.

Figure 6. The sample-reagent pickup mechanism in the resting or buffer position.

positioner, the pipette mechanism is driven by a stepping motor that utilizes optical sensing for precise positioning. It is also computer-controlled, utilizing special check circuits to alert the computer of improper performance.

The reagent block, which may be seen in Figure 5 or 6, is the passive component in the sampling system. It holds the receptacles containing the radioligand and the antiserum solutions and provides a trough through which buffer flows to be picked up when the three probes are in the resting position.

All fluid motion in the Gammaflo^{TM} instrument is provided by a peristaltic pump. All 13 channels of this pump are utilized in normal instrument operation. A single roller assembly whose axis is in the horizontal plane drives the various fluids through the tubing. The bed for each tube channel in the pump is individually adjustable with a specially designed torque adjusting tool to allow for widely differing tubing sizes. Thus a consistent pressure is maintained on each tube, regardless of its diameter, and tube life is considerably enhanced. The computer has the ability to shut down the pump in case of an instrument malfunction, thus preventing contamination. At the end of a normal assay sequence, the roller assembly is automatically disengaged, thereby conserving reagents and increasing tube life.

The individual aliquots of sample, antibody, and radioligand meet at the mixing tee, shown in Figure 7 and located after the peristaltic pump. The sample-reagent mixture is then almost immediately combined with incubation buffer and air.

Incubation takes place in the coils between the bubbler and valve 1. Special fittings have been employed between the tygon tubing and the glass fittings and coils to provide a smooth transition for the reaction mixture so as to reduce lateral diffusion. Two coils, providng an incubation time of about 6 min, are presently employed in the Gammaflo^{TM} instrument. All assays developed to date can be performed at ambient laboratory temperature and are not affected by moderate temperature fluctuations. Some assays under development have shown improved performance when run at higher temperatures. These assays employ a custom-designed, noncirculating liquid bath which provides a nominal incubation temperature of 37 ± 1°C. The controller components and the incubation coils are enclosed in a cylindrical, transparent housing mounted to the right-hand wall of the sink. All electrical connections to the temperature controller are low voltage to prevent shock hazard. A separate, independent temperature sensing element is included in the incubator and is monitored by the computer.

The Gammaflo^{TM} concept necessitates six valves: four 3-way and two 2-way. The proper configuration is attained by using ten pinch valves. Eight of these are utilized in pairs to construct the 3-way valves, valves 1, 2, 5, and 6; a typical 3-way valve is pictured in Figure 8. Each pair of pinch valves is wired in parallel and simultaneously actuated by the computer so that the two members of the pair are always in opposite states, one being open when the other is closed. To illustrate this more fully, let us examine the operation of valve 1. The flow diagram (see Figure 2) shows the incubation mixture flowing to either the sample detection loop or to waste. The tubing path from the incubation coils is bifurcated through a T-fitting with each of the two resulting fluid paths controlled by a pinch valve. The path leading to the sample detection loop is controlled by a pinch valve that is normally open, and the path leading to waste is controlled by a pinch valve that is normally closed. Thus, in the unactuated state, flow to waste is prevented, and the incoming reaction mixture is routed to the sample detection loop. In the actuated state, the pinch valve controlling the path to the sample detection loop is closed and the pinch valve on the waste line is opened allowing the intersample buffer to be sent to waste. This configuraion may seem overly complex in view of the availability of several 3-way valves. However, when viewed in the light of

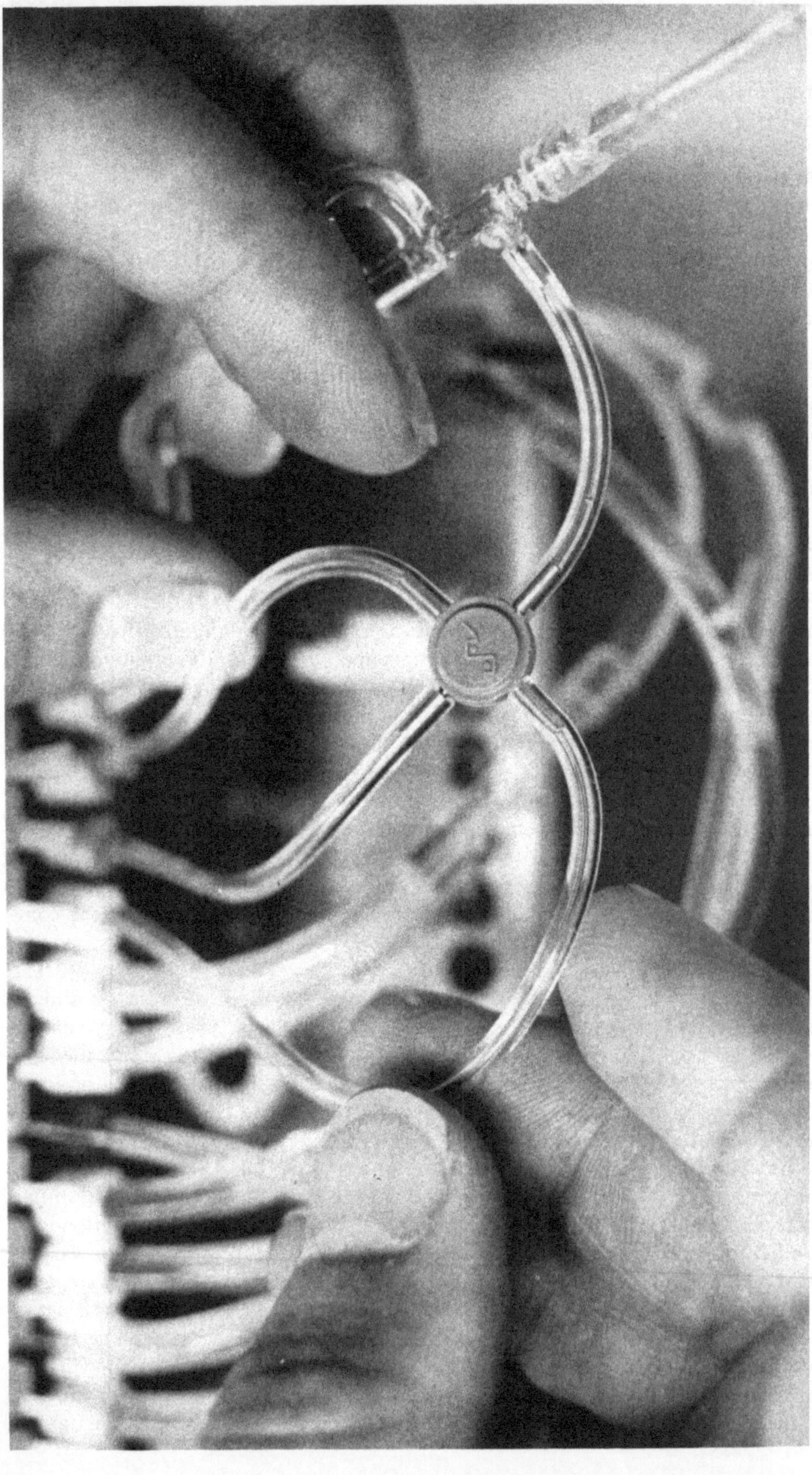

Figure 7. The mixing tee.

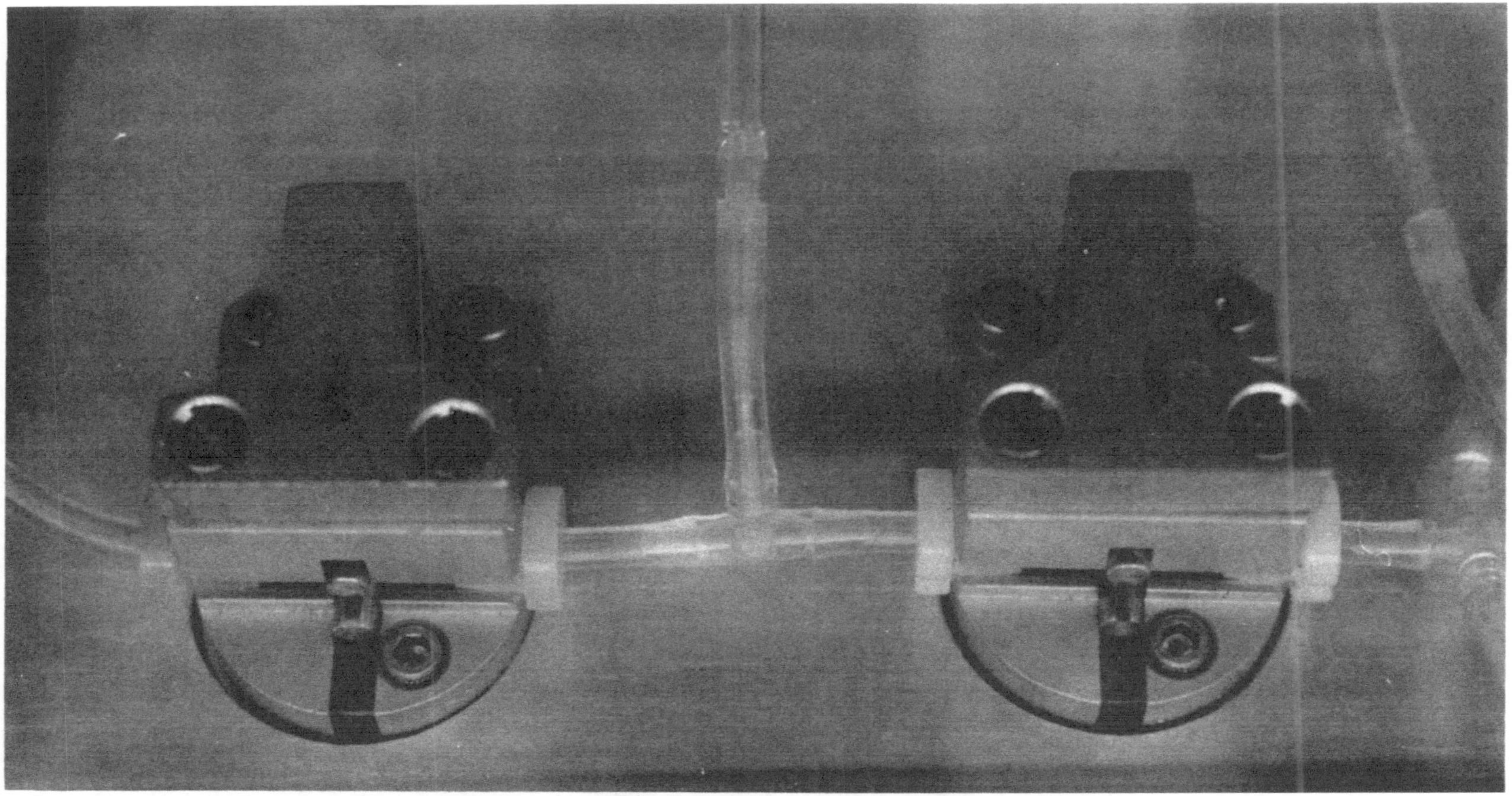

Figure 8. A 3-way valve consisting of two pinch valves and a T-fitting.

reliability and serviceability, pinch valves were found to be superior. If a pinch valve becomes clogged or contaminated, it can be returned to service by cleaning or replacing the tubing. More conventional valves may require complicated disassembly or replacement. Each pinch valve is mounted to the instrument with two easily removable screws and two slip-on electrical connectors. Due to the simplicity of these connections, some users may wish to maintain one or two spare pinch valves in order to perform their own replacement, thus reducing downtime.

Two additional features have been incorporated in the pinch valve to improve its longevity. First, a special silicon tubing which is virtually free from the cold flow problems encountered with other elastic tubings is used. Second, an O-ring is placed around the tubing at the pinch point providing additional force to reopen the tube should some contaminant cause the walls of the tubing to adhere when pinched.

The disposable separation column used with the Gammaflo™ System is made of transparent polypropylene. Its base is press fitted into a specially-designed T-fitting that is part of valve 5. An airtight teflon cap containing the necessary tube connections snaps into the top of the column.

The nuclear counting subsystem consists of a high voltage power supply, a sodium iodide crystal, a photomultiplier tube, a preamp, an amplifier, a single channel analyzer, a counting well, and a computer interface. The computer controls the onset and termination of counting and, of course, reads the results from a register in the interface.

The counting well, which is shown in Figure 9, provides both dynamic and static counting capabilities. The fluids passing through the detection loop are monitored for the appearance of the next reaction mixture. A radioactivity level above the baseline triggers the valve actuation sequence used in the sample separation procedure. At the appropriate time in the sample separation sequence, valve 5 actuates and the antibody-bound radioligand flows from the separation column into a specially designed well inside the sodium iodide counting crystal. Once this container is filled, valve 5 closes and counting begins. When the count is complete, valve 3 opens exposing the material in the counting well to a vacuum created by peristaltic pump. The fluid in the counting well is evacuated and sent to waste.

Special care has been taken to design the instrument in such a way as to reduce its vulnerability to poor power line conditions. All power coming into the instrument first passes through an isolation transformer to reduce common and normal mode noise. DC power is provided by a switching power supply that is capable of producing proper DC voltages with incoming power as low as 90 V. Because the peristaltic pump utilizes a synchronous motor, it can tolerate comparable low-line conditions. An incoming power meter has been built into the power distribution panel to aid the operator in diagnosing gross power line problems. The entire instrument is protected by a ground fault interrupt circuit breaker that is located after the isolation transformer to further eliminate shock hazard.

Fluid level sensors have been implemented in the lower right-hand section of the instrument to prevent run-dry conditions of the buffer or column wash reagents and overflow of the waste container.

All components utilized in the analysis of samples are located in the upper right-hand section of the instrument. They are mounted so that any spillage of fluids will collect in the one-piece fiberglass sink that is equipped with a bottom drain which flows directly into the waste container. Thus should a spill occur, it may be washed down the drain with a minimum of user inconvenience.

At the heart of the instrument is a Digital Equipment Corporation LSI-11® microcomputer that controls all instrument functions and monitors instrument performance.

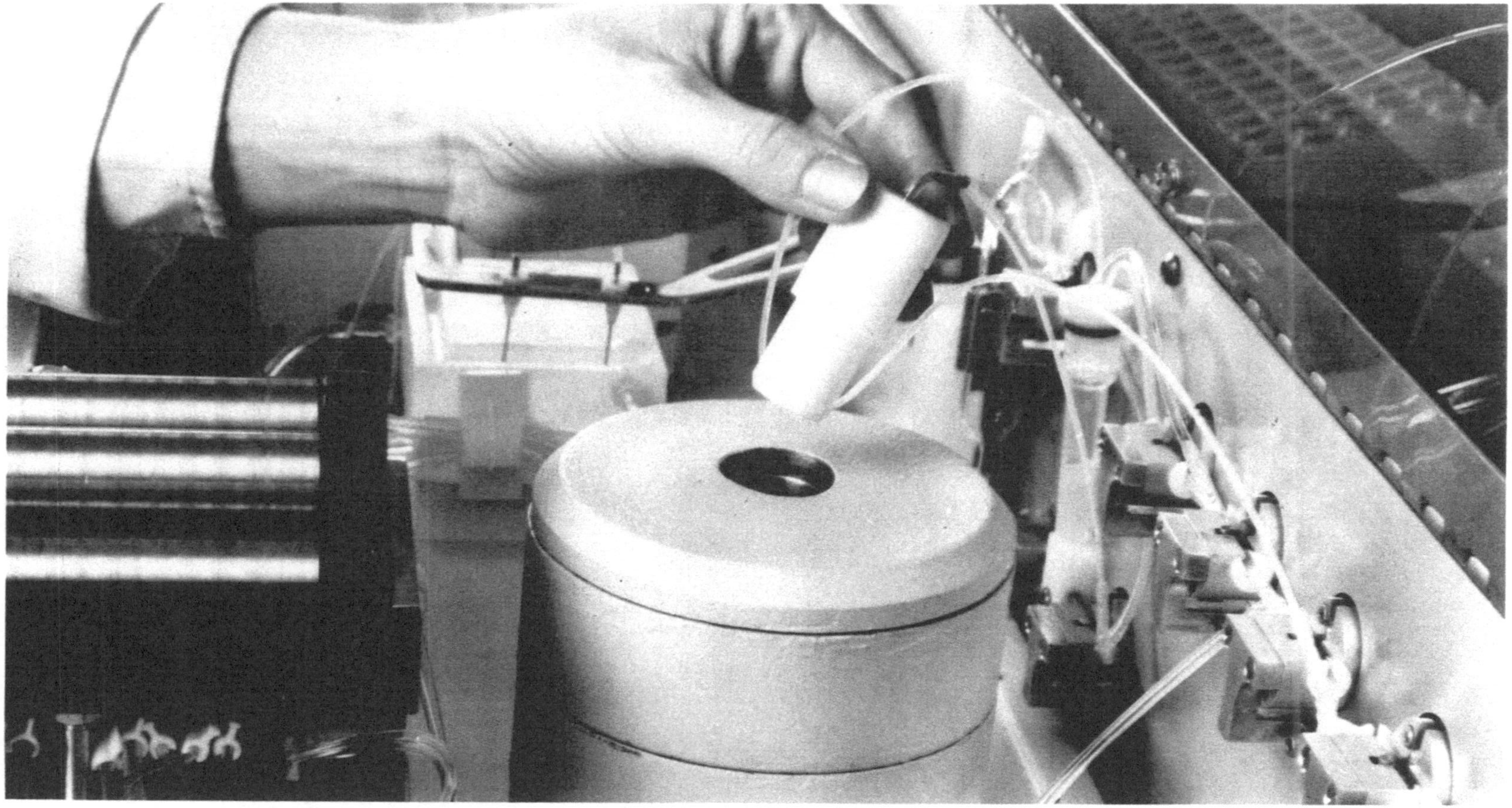

Figure 9. The counting well withdrawn from the gamma counter exposing the sample detection loop.

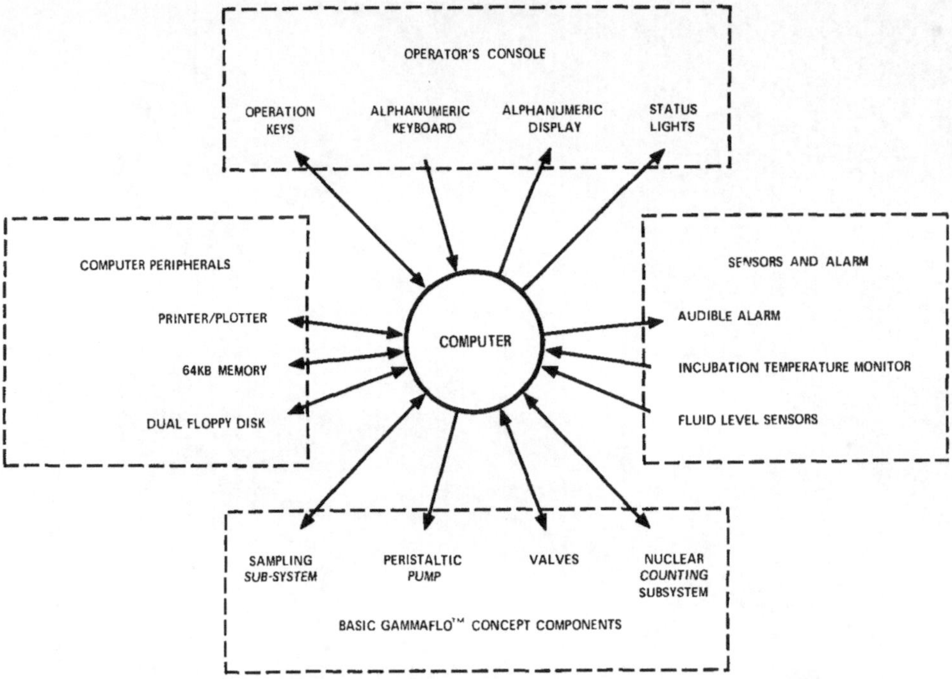

Figure 10. A block diagram of the implementation of the Gammaflo ® concept.

Figure 10 is a block diagram of the Gammaflo^{TM} instrument implementation. Among the features offered by this implementation are an advanced instruction set, hardware floating point multiply and divide instructions, 64,000 bytes of memory, and the ability for software development on a larger scale computer system without translation. Coupled to this computer configuration is a dual floppy disk system, an advanced printer/plotter, and an operator console. This extremely powerful and flexible configuration is superior to hard wired or firmware control and processing circuitry offered in most other clinical instruments. The software being developed for the Gammaflo^{TM} instrument runs under the Digital Equipment Corporation RT-11® operating system. This reliable operating system offers both general file handling support and overlay support.

All printed reports and standard curve graphs are prepared by the instrument's printer/plotter. This impact printer is equipped with a daisy wheel that yields fully formed characters. Standard 11 in. × 14 7/8 in. computer paper is used. In the plotting mode the unit provides medium resolution plots of the standard curve as shown in Figure 11. Horizontal plotting resolution is 120 positions/in., and vertical resolution is 96 positions/in. Plotting performance is further enhanced by a feedback mechanism that assures proper vertical motion.

All components utilized in the Gammaflo^{TM} instrument have been selected and tested to provide minimum downtime. Each finished instrument is thoroughly tested with actual reagents and subjected to an accelerated testing program to detect infant mortality problems. Factory-trained specialists install the Gammaflo^{TM} System and perform thorough testing at the customer's site to further assure satisfactory performance.

V. THE GAMMAFLO^{TM} SYSTEM — OPERATION

The entire operation of Gammaflo^{TM} is under the control of a microcomputer. Upon

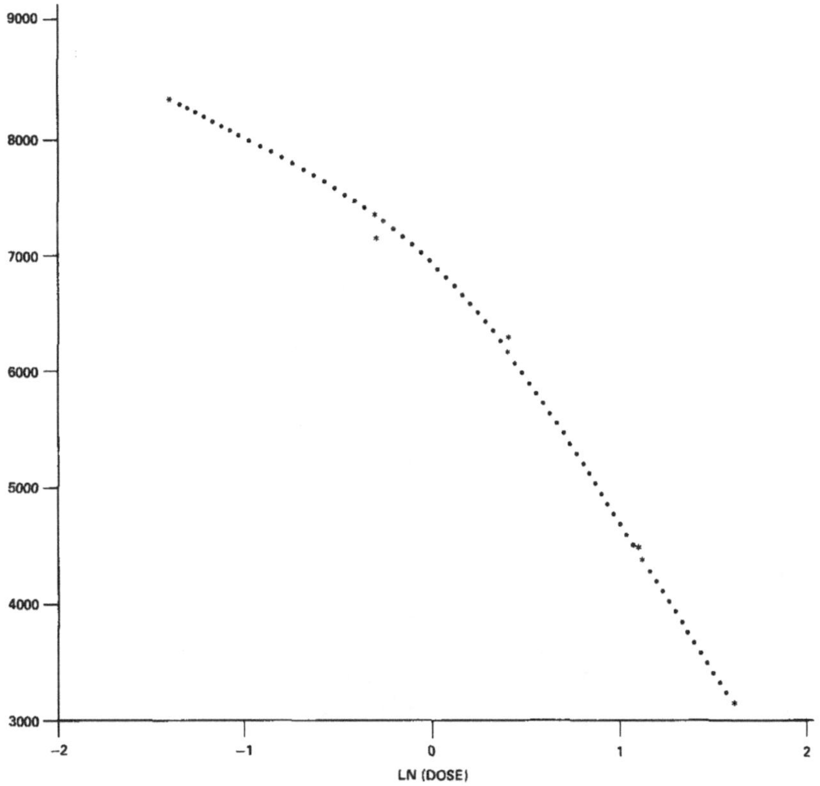

Figure 11. The medium resolution plot of a typical standard curve. Horizontal plotting resolution is 120 positions per inch. and vertical resolution is 96 positions per inch.

start-up, the program (which is supplied and maintained on a floppy disk) is automatically transferred from the disk into the memory of the computer, an operation taking less than 10 sec. It is this copy that actually controls instrument operation.

The operator interfaces with the system via the operator's console which is pictured in Figure 12 and which includes the status lights, the 32 character display, the keyboard, and the eight operations keys. The status lights indicate the current state of the instrument. The display exhibits alphanumeric information entered by the computer or by the operator via the keyboard. The operator utilizes the keyboard and the operations keys to program assays.

A. Functions of Operations Keys

The functions of the individual operations keys are as follows:

1. Select

This key serves two interface roles: first, it transfers user-entered data from the display to the computer; second, it acts as a "yes" key indicating that the operator wishes to utilize the displayed option.

2. Pass

This key is the "no" key and its use has the effect of rejecting the displayed option.

3. Options

This key provides access to a variety of specialized features that may be utilized when the system is not performing an assay.

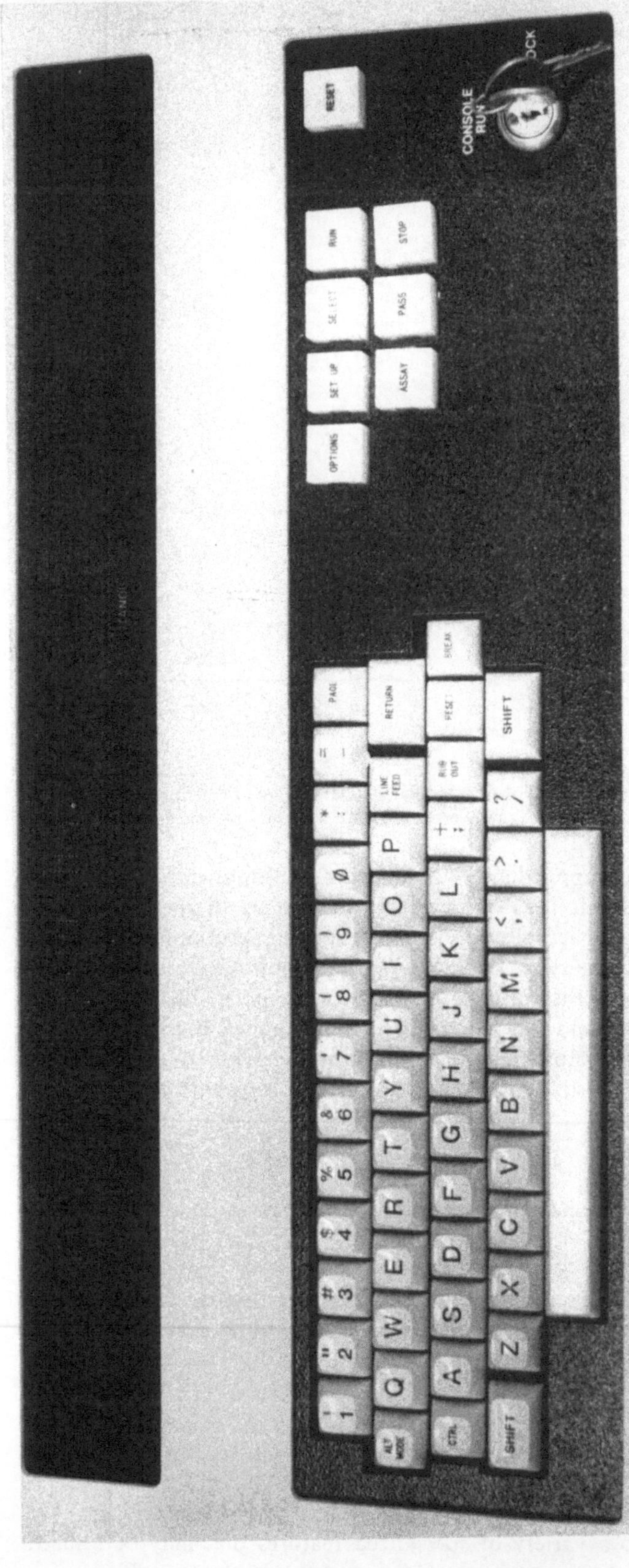

Figure 12. A close-up view of the operator's console including the display, the keyboard, the eight operations keys and the console lock. The display reads enter time (HH:MM) and the standby status light is lit.

4. Setup

This key presents a checklist of operations that the user should complete before performing an assay. Use of this key is mandatory after power up and after an abnormal termination of an assay.

5. Assay

Through the use of this key, the operator programs the instrument for the chemical and mathematical analyses of a particular array of samples.

6. Stop

When the instrument is in the midst of an assay, use of the STOP key prohibits further sample pickup but does not affect the analysis of samples in progress. (Contrast this result with the effect of using the RESET key under the same circumstances — *vida infra*).

7. Run

After the setup procedure has been completed and the instrument has been programmed, depressing this key starts the assay. If the STOP key has been utilized to interrupt the sampling sequence, use of the RUN key causes the pipet mechanism to resume operation where it left off.

8. Reset

This key is always active, having the general effect of terminating the procedure in progress and returning the system to its resting state. However, the specific effect depends upon the status of the system when RESET is pressed, for example:

1. If the instrument is in the midst of the OPTIONS or SETUP modes, the mode is terminated and the OPTIONS and SETUP keys become lit.
2. If the instrument is in the midst of the programming procedure dictated by the ASSAY key, any information or commands already entered are erased and the OPTIONS, SETUP and ASSAY keys become lit.
3. If the instrument is in the midst of an assay, all chemical and mathematical analyses are terminated. The printer stops and a clean sheet of paper is positioned. The valves return to their respective resting state and any samples in progress are ignored.

The computer will respond to legal commands only. Only lit operations keys are active, pressing an unlit key has no effect on instrument operation. Requested alphanumeric information must be entered according to the format shown on the display. All entries must be terminated by depressing the SELECT key. If an error is detected before the SELECT key has been depressed, the user may clear the display by either repeatedly pressing the space bar or by simply pressing the CTRL and I keys simultaneously. Except in the case of entering sample identifications, if an error is not detected until after the SELECT key has been depressed, the user must depress the RESET key and reenter all of the necessary information.

B. Programming for an Assay

After the power is turned on, it takes approximately 10 sec for the operating program to be booted from storage on the floppy disk into the memory of the computer. Upon completion of the process the display requests the user to enter the time and date, thereby setting the computer's internal clock. Next the operator sets up the instrument by engaging the pump and by loading the necessary reagents and samples. Once all of the preliminaries have been completed, an operation which takes about 20 min, the instrument is ready to be programmed for the first assay. The user presses the SETUP key and verifies that the preliminaries have been completed. When the ASSAY key is pressed, a sequence of prompts will appear on the display. The operator must respond as will be discussed below in order to successfully perform an assay. (See Figure 13 for a flow chart.)

1. Type Assay

The instrument maintains an alphabetical listing (including legal abbreviations) of all assay protocols that have been developed. This list may be accessed by entering the letter A and then repeatedly depressing the PASS key.

When preparing to perform an assay, the operator may enter either the full name or the legal abbreviation of the desired assay. A particular assay may also be accessed by entering the first letter of the name of the assay. Because more than one assay may start with the same letter, the operator may have to repeatedly depress the PASS key until the desired assay is displayed before pressing SELECT.

If the operator enters the name or abbreviation of an assay that has no associated protocol, the display "Type Assay" will reappear. The operator should enter the letter A and repeatedly depress the Pass key until the display reads "Assay Development". This sequence allows the instrument to be programmed for assay procedures that have been internally developed. After entering the full name of the assay being developed and specifying its legal abbreviation, the operator will atuomatically be directed into the series of displays described in Section V.C.

2. Operator I/D

Up to 16 characters may be entered. If nothing is entered, this line will not appear on the printout.

3. Full Assay

The computer has the capacity to store one set of standard points for any assay that is run using the standard protocol. Upon completion of a given automated assay performed in the standard mode, the new set of standard points automatically displaces the old set. If the SELECT key is depressed, a new standard curve must be generated. If the PASS key is depressed, the set of standard points in storage will be utilized to plot the standard curve.

When using the Manual Assay Computation option (see Section V.D.8) in the standard mode, the user may elect to save these standard points in which case this set of points displaces whatever set is in storage.

When power is shutdown, all sets of points in storage are lost. To save such sets, the option "Save System Image on Floppy Disk" (see Section V.D.10) must be used.

4. Enter Blank Count

If this option is desired, depress SELECT and enter the blank value. Otherwise, press PASS.

5. Reagent Kit Number

Enter the lot number of the kit being used. If nothing is entered, this line will not appear on the printout.

6. Use Standard Protocol

Preprogrammed assays and the associated standard protocols are completely de-

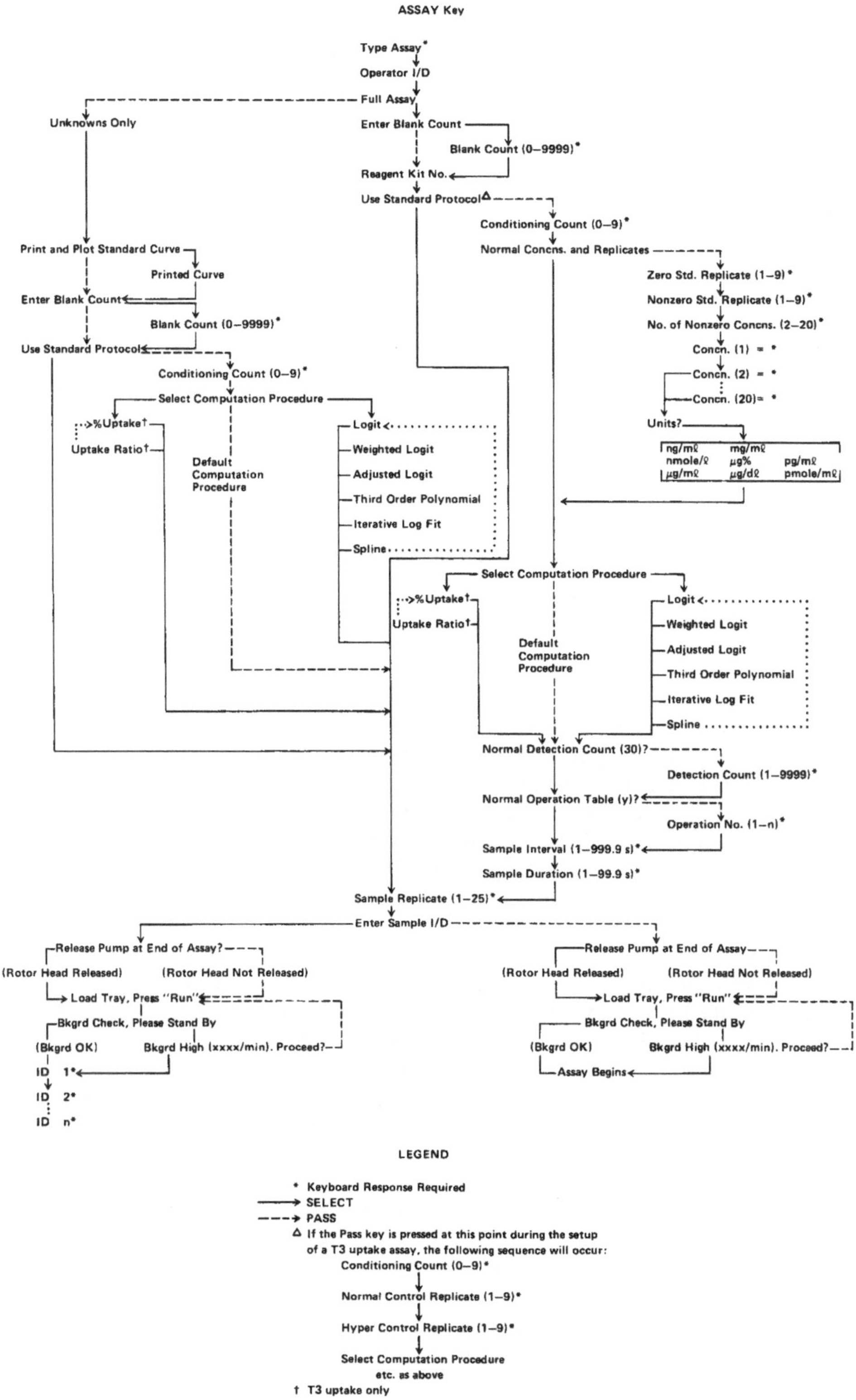

Figure 13. The logic diagram of the assay key.

scribed in the Operations Manual. E. R. Squibb & Sons, Inc. recommends that these protocols be followed for all applicable clinical assays. Therefore, the user should select this option.

7. Sample Replicate (1-25)

The pipet mechanism can be programmed to sample up to 25 times from the same cup. Enter the desired number. Because the entered number will pertain to every unknown, the operator must be sure that each cup contains the necessary amount of sample.

8. Enter Sample I/D

If the PASS key is depressed, no sample identification other than tube number and sample number will appear on the printout. If the SELECT key is depressed, the user will be given the opportunity to enter individual sample identifications after the assay sequence has commenced.

9. Release Pump at End of Assay?

This display asks whether the rotor head should be automatically released from the pump at the completion of an assay. Release of the rotor head conserves buffer and maintains the useful life of the pump tubing. If another assay is planned within an hour after the completion of the first one, then use of this option is unnecessary and the PASS key should be pressed.

10. Load Tray, Press "RUN"

Pressing the RUN key initiates the assay sequence. The display will read "Bkgrd Check, Please Stand By" and the background radioactivity in the gamma counter will be quantitated for 30 sec. If the count is less than 1100 counts per minute, then the instrument will commence the assay. If the count is greater than 1100 counts per minute, the display will read "Bkgrd High (xxxx/min). Proceed?" and the SELECT and PASS keys will become lit. If the operator decides to ignore the computer's warning and to initiate the assay sequence, then the SELECT key should be pressed. Otherwise, the operator should clean the counting chamber and then depress the PASS key. The display will again read "Load Tray, Press RUN" and the RUN key will become active. Pressing the RUN key begins the background check and, if the background is within limit, the assay sequence. The background count will be factored to conform to the counting time called for in the Operation Table being used and the resulting amount will automatically be subtractetd from each subsequent determination. All count values appearing on the printout are observed count less background.

After completing the background determination, the instrument will request the identification of the individual samples in the defined format (see "Define Sample I/D Fields", Section V.D.11). Depress the SELECT key after typing in each entry. If an error in the input is noted before the SELECT key is depressed, clear the display by depressing the CTRL and I keys simultaneously and enter the correct identification. Once the SELECT key has been depressed, any previous identification within the same field that appears on the display can be accessed by holding the shift key down while repeatedly depressing the < key until the particular identification is reached. This identification can now be corrected as explained earlier. To access a higher numbered identification within the same field that appears on the display, hold the shift key down while repeatedly depressing the > key until the particular identification is reached. To switch from one field to another, repeatedly press the SELECT key until the desired field appears on the display as indicated by its associated prompt. Sample identification information may be entered or corrected until the assay results for the sample are printed.

C. Assay Development

The foregoing description illustrates the simplicity and ease of operating the instrument using the standard protocols. As mentioned earlier, the user has the capability to develop his own assay sequences from the keyboard. To do so, he enters "Assay Development" in response to the display "Type Assay", thereby gaining access to the following series of displays.

1. Conditioning Count (0-9)

The instrument can be programmed to pipet up to nine additional times from the *first* cup. This feature, known as the "Conditioning Count", allows the user to condition the system with a protein-containing sample as part of the assay sequence. The results of reaction mixtures assayed during this phase do not appear on the printout.

2. Zero Standard Replicate (0-9)

The pipet mechanism can sample as many as nine times from the cup containing zero standard. Enter the desired number.

3. Nonzero Standard Replicate (1-9)

The pipet mechanism can sample as many as nine times from each standard-containing cup (only one cup may be used for each standard). Enter the desired number of replicates to be taken from each nonzero standard.

4. Number of Nonzero Concentrations (2-20)

Enter the total number of nonzero standards (maximum of 20) followed by the individual concentrations.

5. Units?

This option elaborates the heading of the column in which the numerical results of an assay are tabulated. Repeatedly press the PASS key until the desired designation is displayed, then press SELECT.

6. Select Computation Procedure

The available procedures are listed in Table 1. Repeatedly depress PASS until the desired procedure is displayed, then depress SELECT.

7. Detection Count (1-9999)

Once an assay has begun, the instrument monitors the radioactivity in the gamma counter every second looking for the next sample. The detection count is one of the variables in the algorithim that determines when a reaction mixture has arrived. An entry of 30 has been found to yield optimal instrument operation in all cases examined so far.

8. Select Operation Table

The valve sequence for each assay having a standard protocol is compiled in an "Operation Table". The operator also may have constructed additional operation tables by using the option "Edit Operation Table" (see Section V.D.7). The number of the appropriate table should be entered.

Table 1
AVAILABLE ANALYTICAL PROCEDURES FOR COMPUTATION OF THE STANDARD CURVE

Logit-log
Weighted logit-log
Adjusted end-point logit-log
Iterative log fit
Third-order polynomial
Spline under tension

9. Sample Interval (1—999.9 s)

The length of time, in seconds, from the beginning of one sample to the beginning of the next sample is defined as the sample interval. Enter the desired time to the nearest tenth of a second.

10. Sample Duration (1-99.9 s)

This term is defined as the number of seconds from the point at which the pipet arm leaves the wash wells to the point at which the arm leaves the reagent wells. By definition, this time period is included in the sample interval. Enter the desired time to the nearest tenth of a second.

A typical print out containing assay results is shown in Figures 14 through 17. As can be seen in these figures, the printout includes raw data for the standard curve (Figure 14), standard curve analytical data such as B/B_o values, correlation coefficient, etc. (Figure 15), a graphic plot of the standard curve (Figure 16) and interpolated values for unknown samples (Figure 17).

D. Options

Depressing the OPTIONS key allows access to a variety of specialized features that may be utilized when the system is not performing an assay. Press the PASS key until the desired operation is displayed and then press SELECT. If the desired option is accidentally missed, repeatedly depress the PASS key until it reappears. Use of the RESET key terminates the options sequence (see Figure 18 for a flow diagram).

1. Report Error Conditions

Selection of this option is mandatory at the completion of a run during which the fault signal was activated. Once this option has been chosen, the SELECT key should be repeatedly depressed until all of the error messages have been displayed and the fault signal terminates. No new assay or procedure can be performed while the fault signal is on.

The possible error messages are

1. False triggering — the instrument constantly compares the number of pipet arm operations with the number of separation sequence triggers. If the latter exceeds the former, the fault signal is sounded and the assay sequence halts.
2. Floppy disk write error — This error message is possible only when the option "Save System Image on Floppy Disk" is being used. The most likely causes are a defective disk, an empty drive receptacle, or a disk that is write-protected.

GAMMAFLO™ STANDARD COUNTS

CORTISOL 12:52:28 15-JAN-79

REAGENT KIT NO. B2303 RGNTS
OPERATOR I/D CLINICAL B

SEQ. NO.	TUBE NO.	COUNT	CONC. (UG/DL)	
1	1	10048	0	(05:30-1278-25-15-47.0)
2		10138	0	(05:29-1231-24-14-47.0)
3	2	9272	1.00	(05:30-1307-26-16-47.0)
4		9182	1.00	(05:30-1264-27-15-47.0)
5	3	6817	5.00	(05:30-1270-24-15-47.0)
6		6853	5.00	(05:31-1096-27-16-47.0)
7	4	5391	10.00	(05:30-1150-26-14-47.0)
8		5584	10.00	(05:30-1069-24-15-47.0)
9	5	4217	20.00	(05:31-1096-28-16-47.0)
10		4290	20.00	(05:31-1104-23-14-47.0)
11	6	2910	50.00	(05:30-1066-22-14-47.0)
12		2850	50.00	(05:30-1089-21-15-47.0)

Figure 14. Raw data for computation of a standard curve.

GAMMAFLO™ STANDARD CURVE DATA

CORTISOL 12:52:28 15-JAN-79

REAGENT KIT NO. B23D3 RGNTS
OPERATOR I/D CLINICAL B

SEQ. NO.	TUBE NO.	COUNT	B/BO	CONC. (UG/DL)	CALCD.	% DIFFCE.
1	1	10048		0		
2		10138		0		
3	2	9272	0.919	1.00	<1.00	
4		9182	0.910	1.00	1.04	4.3
5	3	6817	0.675	5.00	4.99	-0.2
6		6853	0.679	5.00	4.89	-2.1
7	4	5391	0.534	10.00	10.66	6.6
8		5584	0.553	10.00	9.63	-3.7
9	5	4217	0.418	20.00	20.32	1.6
10		4290	0.425	20.00	19.47	-2.7
11	6	2910	0.288	50.00	48.98	-2.0
12		2850	0.282	50.00	>50.00	

THIRD-ORDER POLYNOMIAL FIT OF (COUNT) VS (SQRT(DOSE))
A = 11876.2 B = -2990.6 C = 362.9 D = -17.0
CORRELATION COEFF. = 0.9997

Figure 15. Computation and analysis of raw data.

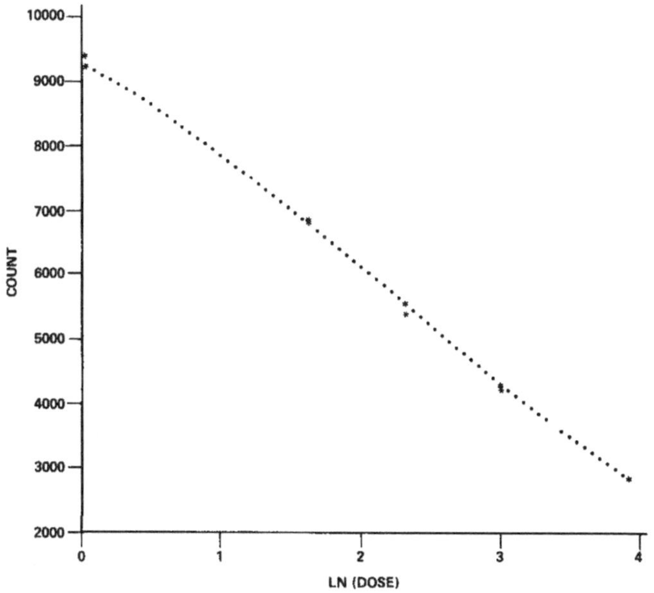

Figure 16. The standard curve derived from the data shown in Figure 14 and the analysis shown in Figure 15.

GAMMAFLO™ INTERPOLATED SAMPLES

CORTISOL 12:52:28 15-JAN-79

REAGENT KIT NO. B23D3 RGNTS
OPERATOR I/D CLINICAL B

SEQ. NO.	TUBE NO.	ID	COUNT	B/BO	CONC. (UG/DL)	MEAN	
13	7	DADE 110	7164	0.710	4.11		(05:30-1134-21-15-47.0)
14			6972	0.691	4.58	4.34	(05:30-1198-23-15-47.0)
15	8	DADE 210	4607	0.456	16.26		(05:29-1272-24-14-47.0)
16			4633	0.459	16.02	16.14	(05:29-1140-22-15-47.0)
17	9	15	5297	0.525	11.20		(05:30-1203-24-16-47.0)
18			5313	0.526	11.11	11.15	(05:30-1194-24-15-47.0)
19	10	15B	5605	0.555	9.53		(05:30-1278-23-15-47.0)
20			5650	0.560	9.30	9.41	(05:31-1137-26-15-47.0)
21	11	15 C	5879	0.582	8.25		(05:30-1240-23-14-47.0)
22			5902	0.585	8.15	8.20	(05:30-1227-25-15-47.0)
23	12	16	5606	0.555	9.52		(05:31-1223-27-16-47.0)
24			5506	0.546	10.03	9.78	(05:30-1230-25-14-47.0)
25	13	16B	5483	0.543	10.16		(05:30-1249-25-15-47.0)
26			5492	0.544	10.11	10.13	(05:29-1323-25-14-47.0)
27	14	16C	5785	0.573	8.67		(05:30-1242-28-16-47.0)
28			5819	0.577	8.51	8.59	(05:29-1302-25-14-47.0)
29	15	17	5040	0.499	12.84		(05:29-1247-26-15-47.0)
30			4986	0.494	13.22	13.03	(05:30-1301-27-15-47.0)
31	16	17A	6369	0.631	6.37		(05:30-1317-25-15-47.0)
32			6514	0.645	5.89	6.13	(05:31-1201-24-16-47.0)
33	17	17AH	6549	0.649	5.78		(05:30-1287-27-14-47.0)
34			6780	0.672	5.10	5.44	(05:30-1290-25-15-47.0)
35	18	17D	5024	0.498	12.95		(05:29-1247-25-14-47.0)
36			4850	0.481	14.22	13.59	(05:30-1210-24-16-47.0)
37	19	17E	5405	0.536	10.58		(05:29-1303-24-14-47.0)
38			5419	0.537	10.50	10.54	(05:30-1255-25-16-47.0)
39	20	18	4929	0.488	13.63		(05:29-1226-21-14-47.0)
40			4857	0.481	14.17	13.90	(05:30-1266-24-15-47.0)
41	21	18A	6035	0.598	7.60		(05:30-1208-23-15-47.0)
42			5970	0.591	7.86	7.73	(05:30-1230-23-15-47.0)
43	22	18AH	6068	0.601	7.47		(05:30-1306-26-15-47.0)
44			5982	0.593	7.81	7.64	(05:31-1253-28-16-47.0)
45	23	18D	5040	0.499	12.84		(05:30-1299-24-14-47.0)
46			4972	0.493	13.32	13.08	(05:29-1325-24-14-47.0)
47	24	18E	5176	0.513	11.94		(05:29-1257-21-15-47.0)
48			5385	0.534	10.69	11.32	(05:30-1236-27-16-47.0)
49	25	DADE 110	7251	0.718	3.91		(05:29-1265-25-14-47.0)
50			7014	0.695	4.47	4.19	(05:31-1234-25-16-47.0)
51	26	DADE 210	4717	0.467	15.30		(05:29-1316-26-13-47.0)
52			4868	0.482	14.09	14.69	(05:29-1332-27-15-47.0)
53	27	19	4747	0.470	15.05		(05:31-1281-26-17-47.0)
54			4790	0.475	14.70	14.87	(05:29-1353-25-13-47.0)
55	28	19A	7539	0.747	3.30		(05:30-1305-25-16-47.0)
56			7705	0.763	2.99	3.14	(05:29-1406-27-14-47.0)
57	29	19D	5009	0.496	13.06		(05:30-1286-30-16-47.0)
58			5079	0.503	12.58	12.82	(05:31-1272-26-16-47.0)

Figure 17. Printout of unknown values interpolated against the standard curve depicted in Figure 16.

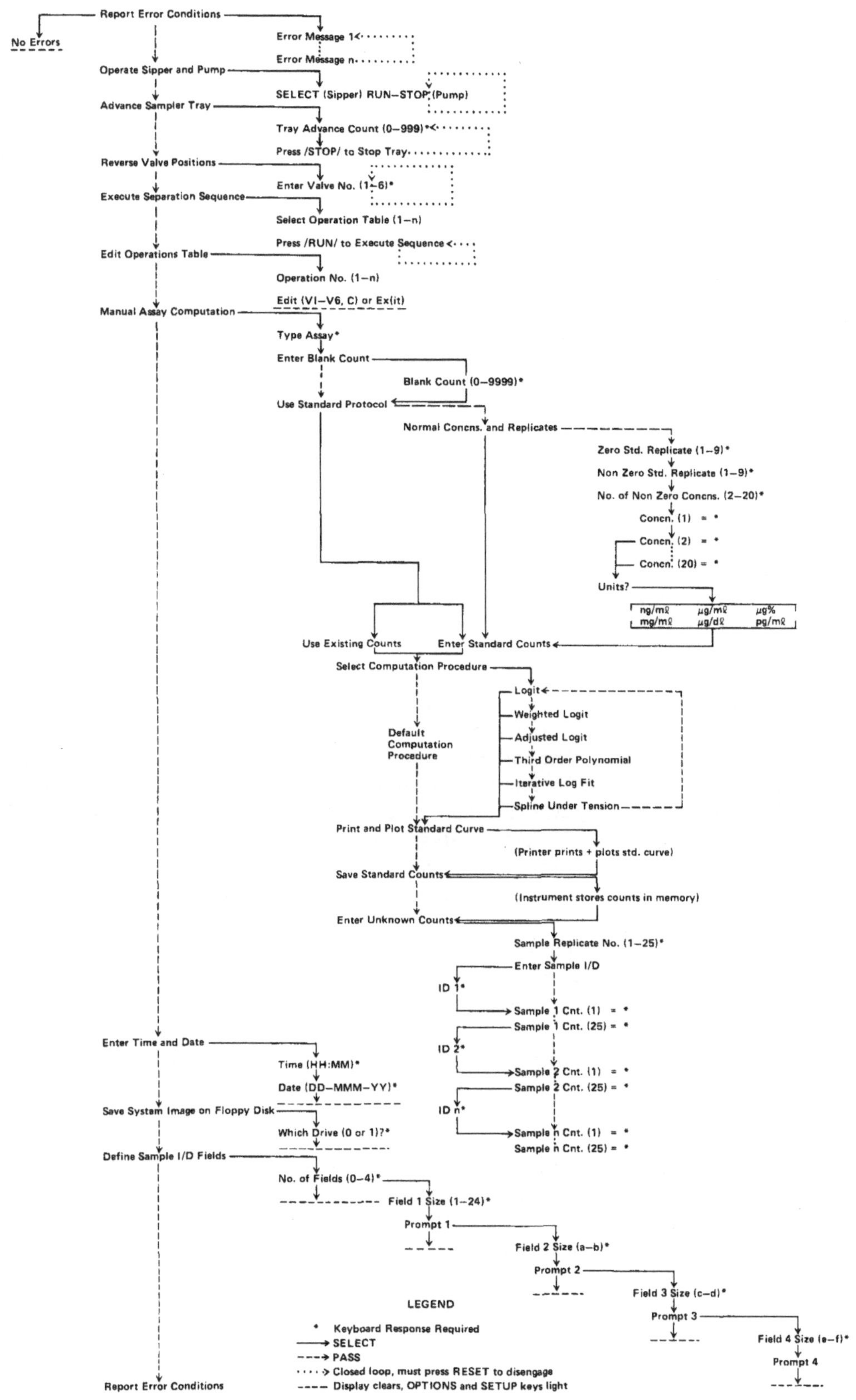

Figure 18. The logic diagram of the options key.

3. Incubation time error — The instrument continuously monitors the time interval from sample pickup to activation of the valve sequence. The error response to an untimely activation depends on the actual arrival of the reaction mixture.
4. Out of buffer — Each reservoir sits on a weight sensor. When the amount of liquid present falls below a given level, the sensor is tripped causing activation of the fault signal and termination of the assay procedure.
5. Out of column wash — See 4. Out of buffer.
6. Out of paper — The printer is equipped with an optical device that senses when the paper supply is exhausted. When the sensor is tripped, the fault signal activates and the assay procedure terminates.
7. Pipet sense error — Should the pipet mechanism fail to operate upon computer command, the fault signal is activated and the assay sequence is terminated. Because an interlock prevents operation of the pipet arm while the tray is in motion, an error message is indicative of either a malfunctioning pipet arm or of a moving tray.
8. Printer error — Should there be a printer malfunction that is not self-correctable, the fault signal will sound and the assay procedure terminates.
9. Radioactive waste full — The waste container also sits on a weight sensor. When the radioactive waste container is full, the fault signal activates and the assay procedures terminates.
10. Sample queue overflow — Each pipet arm operation causes an entry to be made in the memory of the computer. As each reaction mixture is assayed, the oldest entry is removed. Space has been allotted for 20 in-progress determinations.
11. Temperature sensor error — If there is a hardware malfunction in the analog-to-digital temperature converting circuitry, then the fault signal will light but the assay will continue. This error message can occur only if the customized option "Incubation Time Print" is in use.
12. Tray advance error — The computer continuously monitors tray movement during an assay procedure. If the tray fails to move upon command, the fault signal will light but the assay will continue. The computer will issue up to three additional commands in an effort to advance the tray. If the malfunction persists, the fault alarm will sound and the assay sequence will terminate.
13. Valve error — If a valve cannot cycle it is considered to be in the wrong position. The fault signal sounds and the assay sequence terminates.
14. Valve reset timeout — During certain phases of operation, the instrument automatically returns the valves to their resting state. If this cannot be done within a reasonable time, then the fault alarm will sound.

If this option is elected and there are no error messages, the display will read "No Errors" and the OPTIONS and SETUP keys will become lit.

2. Operate Pipet and Pump

When the SELECT key is pressed, the pipet arm moves from its resting configuration (as shown in Figure 6) and remains in the pipetting position (as shown in Figure 5) until the key is pressed a second time, after which the arm returns to the wash well. When the STOP key is depressed, the pump stops and is restarted by depressing the RUN key. The RESET key must be used to escape this option.

3. Advance Sample Tray

The operator may enter the exact number of positions to be advanced. If a tray advance count of zero is entered, the tray will continue to rotate until STOP is pressed.

The STOP key overrides any entry causing the advance mechanism to halt. Upon completion of any movement, the display will again read Tray Advance Count (0-999). Therefore, the RESET key must be used to escape this option.

4. Reverse Valve Positions

The sense of each solenoid pinch valve has been preset. This option allows the user to temporarily reverse the sense of a valve: a two-way valve that is normally open becomes closed and vice versa; with a three-way valve, the normally open half is closed and the normally closed half is opened — neither half can be operated independently.

When this option is chosen the display will read "Enter Valve No. (1-6)" asking the user to enter the number of the valve to be reversed. The display will transiently show a series of dashes and then the message "Enter Valve No. (1-6)" will reappear signifying the completion of the reversal. The operator may now switch the same valve or any other valve. There is no limit on the number of times that a valve may be reversed. The user must press the RESET key to escape this option. Doing so automatically returns all of the valves to their preset positions. Therefore, the preset position of any valve cannot be permanently changed.

5. Execute Separation Sequence

This option artifically activates the separation sequence and is used when breaking in a new column. After this option has been chosen by depression of the SELECT key, the display will read "Select Operation Table (1-n)". Enter the number of the operations table corresponding to the forthcoming assay. Depression of the RUN key will then trigger the sequence. Depression of the RESET key terminates any sequence in progress returning the valves to their resting configuration and causing the OPTIONS and SETUP keys to become active.

6. Gamma Counter Option

This option allows the instrument to be used as a gamma counter. The counting chamber should be removed from the well and replaced with the sample. The first display, "Count Time (1-999 secs) = ", requests the operator enter the counting time. Pressing the RUN key starts the counter. The running count will be shown on the digital counter (the counter lacks a units digit). At the end of the specified period the actual count will appear on the display. The user may press RUN again to recount the same sample or to count another sample. Before initiating an assay sequence the counting chamber should be replaced being careful not to pinch the telfon tubing. It is not recommended that the sample be pipetted directly into the counting chamber because of the possibility of contamination. However, the top may be removed from the counting chamber and the tube ccontaining the sample may be placed in the base.

7. Edit Operation Table

Each assay that has a standard protocol has an established valve timing sequence that has been compiled into an "Operation Table". This option allows the user to change any of the individual valve timings in a given table or to construct a new table.

The first display, "Operation No. (1-n)", where $n < 9$, asks for the number of the table to be modified. The nth table, except when $n = 9$, is always empty. To determine the status of the ninth table, the user should enter "9" and then "EX". If the table is empty, the display will so state. If the table is in use, the display will clear and the OPTIONS and SETUP keys will become active. Because these tables have numerical designations only, it is important for the user to maintain an accurate record of added or modified tables.

The second display reads "Edit (V1-V6, C) or Ex(it), where C represents the gamma counter. When the user enters either "V" and the number of the valve or "C", the display will indicate the existing timing sequence in one of the following formats:

O—aaC—bbO

or

O—aaC—bbO—xxC—yyO

Where O = Open; C = Closed; aa, bb, xx, yy = Time in seconds after trigger.

Where O = Open; C = Closed; aa, bb, xx, yy = Time in seconds after trigger.

depending upon whether the particular valve cycles once or twice during each separation sequence. Any new timing sequence must be entered in the proper format and must meet the following requirements:

1. The sequence must begin and end with the letter "O".
2. "O"'s and "C"'s must alternate.
3. The times must be in ascending order to the nearest tenth of a second.
4. No valve may cycle (O-C-O) in under one second.
5. No valve or group of valves may cycle less tha 0.6 sec after any other valve or group of valves; however, all six valves may be cycled simultaneously.

To change the valve timing the user first depresses the CTRL and I keys simultaneously, thus clearing the display. The new sequence is then entered and the Select key is pressed. If the entered sequence does not fulfill the above requirements, then the original valve sequence will reappear on the display. If the entered sequence is acceptable, then the display "Edit (V1-V6, C) or Ex(it)" will reappear. Once all of the desired changes have been made on a given table, the user should enter "EX" thus incorporating these changes into the operating program. However, any such changes will be lost when power is shutdown. In order to make these changes permanent, they must be incorporated on the floppy disk through the use of the option "Save System Image on Floppy Disk" (see Section V.D.10).

8. Manual Assay Computation

The displays and the alternatives within this option are very similar to those within the ASSAY sequence (see Section V.B.). The differences are as follow:

1. The display "Use Existing Counts", which appears in this mode, asks the same question as the display "Full Assay" in the ASSAY mode.
2. The computational procedure is not included in the display "Use Standard Procedure". A separate display "Select Computation Procedure" appears. (see Section V.C.6 for an explanation of its use.)
3. The operator can elect not to have the standard curve printed and plotted.
4. The operator can choose whether or not to save the set of standard points that was manually entered. If this set is saved, it displaces the set already in memory. If the standard protocol was not used, then the standard values cannot be saved and the display "Save Standard Counts" will not appear.
5. If the operator elects to enter individual sample identifications, each identification is entered as required rather than at one time. The format of the sample

identification printout is established through the option "Define Sample I/D Fields" (see Section V.D.11). If the PASS key is depressed after the display "Enter Unknown Counts", then the display clears and the OPTIONS and SETUP keys become active.

9. Enter Time and Date

Because the instrument is equipped with a 24-hr clock, the user should convert 12-hr time to 24-hr time. For example, 7:15 am = 07:15 and 7:15 pm = 19:15.

The legal format for the date consists of two numeric characters for the day of the month, hyphen, the first three alpha characters in the name of the month, hyphen, and two numeric characters for the year. For example, September 1, 1978 = 01-SEP-78.

After the date has been entered, the display will automatically clear and the OPTIONS and SETUP keys will become active.

10. Save System Image on Floppy Disk

The operating program is actually a copy that has been booted from the floppy disk into memory. Under normal circumstances, there is no reverse flow of information from memory back to the floppy disk. This option allows that reverse flow — whatever is in memory *at the moment* this option is exercised is tranferred to the floppy disk, *overriding anything that may already be on the disk.*

The instrument contains two separate floppy disk drive systems. The drive designated by the DRIVE Ø SELECT switch is denoted "Ø", the other being "1". Because the information in memory may be transferred to a disk located in either drive, the instrument will query "Which Drive (Ø or 1)?"

In order for this operation to be successful, the disk must be write-enabled by covering the small notch at its rear edge with opaque tape.

11. Define Sample I/D Fields

The sample identification column of the printout has a fixed width of 32 characters and a fixed title of "ID". With this option, this array can be divided into as many as four titleless subcolumns called "Fields". The selected format will be used for every subsequent assay and manual calculation, irrespective of type. The first display within this option reads "Number of Fields (0-4)". Selection of zero returns the printout to the default format (no subcolumns), the display clears and the OPTIONS and SETUP keys become active. The second display reads "Field 1 Size (1-24)", the field size being the number of spaces actually available for a particular sample identification. The third display, "Prompt 1", requests the subcolumn designation; this recognition factor should be as abbreviated as possible. After the last prompt, the display automatically clears and the OPTIONS and SETUP keys become active.

12. Customization Option

The incubation time for each determination, which is defined as the interval, in minutes and seconds, from sample pickup to activation of the valve sequence, will automatically appear on every printout. Additional diagnostic information, that may be helpful for assay development or quality control, is available through the use of this option. As each possible instrument diagnostic appears on the display, the operator signifies acceptance by depressing the SELECT key or rejection by depressing the PASS key. Selected diagnostics will appear in the order given below, with the incubation time appearing first, on every subsequent printout as part of the information reported for each sample. The operator may change the diagnostics being reported by reaccessing the Customization Option and commanding the instrument accordingly.

1. Incubation temperature print — A temperature sensor is included in the temperature controller.
2. Predetection interval print — This interval is defined as the time in seconds from completion of one separation sequence to trigger of the next sequence.
3. Predetection mean count print — This count is the average number of counts per second observed during the pre-detection interval.
4. Kinetic count print — The radioactivity contained in each reaction mixture triggers the assay sequence as it flows through the teflon tubing on the outside of the counting chamber (the detection loop). With this diagnostic the amount of radioactivity passing through the loop between trigger and operation of the first valve is totaled.

VI. THE GAMMAFLO^{TM} SYSTEM — REAGENT DEVELOPMENT

The main objectives of the Gammaflo^{TM} reagent development program were the use of liquid-phase reagents and readily available separation media. Both goals have been achieved. All of the reagents are formulated in liquid-phase matrices. Common separants, such as ion exchange resins, are used to separate the free and antibody-bound fractions. Gammaflo^{TM} assays do not employ the dedicated reagents that are required by many other automated systems.

Because Gammaflo^{TM} is a continuous flow system, the same column is repetitively utilized to separate the antibody-bound ligand from the free ligand. In the early stages of development, it was found that the lifetime of each column was limited by the accumulation of serum components. However, the institution of a column wash cycle after each separation prolonged the serviceability of each column.

In the development of reagents, it was important that as many features as possible be retained among assay systems thereby minimizing the amount of time needed to change from one assay to another. To date, the same pump tubing configuration is utilized for all assays. The volume of reagents used in each assay procedure is varied by changing the timing sequence. Each assay also uses a standard incubation coil configuration. With respect to reagents, the same column wash solution is used in the digoxin and cortisol assays. All thyroid hormone assays employ common buffer and column wash solutions. The buffer and column wash solutions for all assays are provided in bulk quantitites. The other assay components, the radioligand, antibody, standards and separation column, are provided in a kit sufficient for 200 determinations.

A notable feature of the Gammaflo^{TM} instrument is the relatively brief incubation time of 5 to 7 min. This is made possible by short sampling times that keep incubation volume to a minimum, thus resulting in a highly concentrated reaction mixture. Typical sample and reagent volumes in Gammaflo^{TM} assays vary between 10 to 35 $\mu\ell$. Of course, the quality of the antiserum and the radioligand employed also play an important role in attaining satisfactory assay sensitivity.

Working with the original prototype instrument, Dr. Brooker and his co-workers[1] demonstrated the ease with which radioimmunoassays for compounds of such diverse structure as digoxin, insulin, cyclic AMP, cyclic GMP, angiotensin I, and thyroxine could be adapted to the system. The radioligands and antisera employed were either obtained commercially or prepared according to published procedures.[1] To date further effort has lead to the commercial development of the following assays:

- Cortisol Radioimmunoassay
- Digoxin Radioimmunoassay

Table 2
NECESSARY REAGENTS FOR THE DIGOXIN RADIOIMMUNOASSAY

Antiserum
^{125}I Digoxigenin
Digoxin standard solution, 0.25 ng/mℓ
Digoxin standard solution, 0.75 ng/mℓ
Digoxin standard solution, 1.5 ng/mℓ
Digoxin standard solution, 3.0 ng/mℓ
Digoxin standard solution, 5.0 ng/mℓ
Separation column
Buffer concentrate
Column wash solution

Note: All components exc.pt the last two are supplied in kit form, the amounts being sufficient for 200 determinations. The last two components are supplied in bulk, separately.

- T_3 Uptake Radioassay
- T_3 Radioimmunoassay
- T_4 Radioimmunoassay

Let us look in greater detail at the digoxin radioimmunoassay. A list of the necessary reagents appears in Table 2. Assay parameters are compiled in Table 3.

Reproducibility of digoxin results by the Gammaflo™ System are given below as the coefficient of variation (CV%):

	Digoxin concentration	
Parameter	2.0 ng/mℓ	4.0 ng/mℓ
Intra-assay		
within vial (CV%)	5.16	3.89
among vials (CV%)	3.74	4.13
Inter-assay (CV%)	7.77	5.83

Studies on the recovery from serum of added digoxin have also been conducted with Gammaflo™ over a wide range of concentrations. The recoveries were essentially quantitative.

In 1975 the National Committee for Clinical Laboratory Standards in a document entitled, Protocol for Establishing the Precision and Accuracy of Automated Analytic Systems, called attention to two parameters, carry-over and drift, that are uniquely applicable to automated systems. An estimate of percentage carry-over may be obtained from a sequence of samples $H_1H_2L_1L_2$ where H represents a high sample, 4.0 ng/mℓ of digoxin, and L a low one, 0.5 ng/mℓ. It is assumed that the values observed for H_2 and L_2 are true ones because they are preceded by samples of the same concentration, then percentage carry-over p may be defined by the following equation:

$$p \equiv [(L_1 - L_2) / (H_2 - L_2)] \times 100\% \qquad (1)$$

Table 3
ASSAY PARAMETERS FOR THE DIGOXIN RADIOIMMUNOASSAY

Replicate — all standards are run in duplicate
Conditioning count — 3
Normal detection count — 30
Recommended computation procedure — third-order polynomial
Sample interval — 90 sec
Sample duration — 7.8 sec
Buffer concentrate dilution — 1:10 with distilled water
Operation table number — 1
Valve timings

V1	18—62
V2	18—29, 50—62
V3	62—69
V4	18—62
V5	10—30
V6	30—46
C	30—62

If a particular sample is analyzed twice during an assay sequence, once near the beginning and again 50 tubes later, obtaining values Y_1 and Y_2, then the percentage drift may be defined as

$$d \equiv [(Y_2 - Y_1) / Y_1] \times 100\% \qquad (2)$$

when any effects due to carry-over have been eliminated. Carry-over and drift for the digoxin assay were observed to be not significantly different from zero at the 95% confidence level.

VII. DISCUSSION

The incorporation of the Gammaflo™ concept into a commercially available, computer controlled instrument will automate a wide variety of radioimmunoassays in both the clinical laboratory and the research laboratory. The basic concept is exceptionally versatile because the antibody-antigen reaction occurs in solution and because the separation of free from bound ligand occurs *via* column chromatography. Therefore, the properties of different antibody-antigen incubation mixtures can be accomodated by simply changing separants. Work is in progress on a separation medium that may be applicable to all radioimmunoassays.

From the keyboard of the instrument, an operator can write customized programs, for *any* assay, thereby fulfilling particular needs. This feature does not require any knowledge of computer programming. The storage of the operating program on a floppy disk adds another dimension of flexibility because the insertion of a new disk instantly updates the instrument. The dual drive floppy disk system allows one disk to be used for long term data storage, thereby reducing the amount of paperwork that must be filed.

ACKNOWLEDGMENT

The authors wish to acknowledge the contributions of Drs. Peter Black and Jan Tu and Mr. Ted Altman in the development of the Gammaflo™ Automated Assay System and the assistance of Dr. Philip Freidenreich in the preparation of this manuscript.

REFERENCE

1. **Brooker, G., Terasaki, W. L., and Price, M. G., Gammaflow®**: completely automated radioimmunoassay system, *Science*, 194, 270, 1976.

Chapter 6

AUTOMATION OF RADIOIMMUNOASSAY WITH THE PICKER PACE-4® SYSTEM

M. L. Simons and W. Piknosh

TABLE OF CONTENTS

I. INTRODUCTION

Since their introduction more than 10 years ago, radioimmunoassay (RIA) and competitive binding radioassays have gained wide acceptance and found increasing application in clinical laboratories. Originally developed to measure peptide hormones, the concept has been extended and refined so that a wide variety of substances can now be measured including steroids, drugs, tumor antigens, and viral marker proteins. The growth in both scope and volume of radioimmunoassays has reached the proportions where many laboratories feel the need to automate the performance of these assays.

The procedures developed for RIA all include pipetting of sample and reagents, incubations of the reaction mixture, separation of the bound and free antigen, measurement of the radioactivity in the appropriate fraction, and calculation of the results. Semiautomated equipment has been applied to the sample and reagent dispense steps, and gamma counters and calculators aid in determination of assay results. Yet even with these techniques, RIA still remains a sensitive and labor-intensive technique.

Criteria for development of an automated radioimmunoassay system include improved precision, flexibility to accomodate a wide range of assays, a high throughput rate, and minimal manual intervention.

The Picker Pace-4® system was developed according to these criteria. The Pace-4® is a refinement of the system described by Professor Kenneth D. Bagshawe of Charing Cross Hospital, London.[1] Pace-4® utilizes discrete reaction tubes for performance of the radioimmunoassay followed by filtration of the reaction mixture through a glass fiber filter strip. Radioactivity retained on the filter is measured for determination of analyte concentration.

II. GENERAL DESCRIPTION

The main component of the Pace-4® system, the chemistry processing unit is illustrated in Figure 1. Immunochemical reaction takes place within individual 2.7 m*ℓ* plastic tubes. The tubes are contained within a cassette which holds up to 160 reactions in fixed sequence. Figure 2 illustrates a fully loaded reaction cassette.

Antigen bound to antibody is separated from free antigen at the end of the reaction period by filtration onto discrete glass fiber filters. The particulates are trapped by the filter which is supported on a strip of 35 mm plastic tape. The radioactivity retained on the filter is counted by a series of parallel gamma detectors. Microprocessor controlled electronics analyze the input from the detectors and process the results according to the data reduction routine selected by the operator. The microprocessor controls all system functions and performs quality control analyses on the assay results.

The operating areas of the system — sample transfer, reagent addition, incubation, and filtration/counting — operate independently of each other under microprocessor control so that several assays may be in process at different stages simultaneously.

The assay results are printed as soon as available along with the quality control parameters to provide the operator with a permanent record of results. Interfaces are provided with the Pace-4® system for transferring data into a laboratory computer or general data storage system.

III. SYSTEM OPERATION

To begin an assay on the Pace-4® system, the sample cassette is prepared with the standards and samples in the sequence in which they will be counted. Sufficient volume must be provided for the appropriate amount to be aspirated for the assay, but meas-

FIGURE 1. Picker Pace-4® System.

urement of the volume is not required. The assay protocol is recalled from the library of protocols stored in the system memory and any changes which may be required for an individual assay are made. Information for the assay including cassette number, location of reagents, and sample sequence is entered. Patient information can be entered in a corresponding patient information file and will be printed with the results. All operator instructions and information are entered via the data entry terminal and stored in the system memory.

The cassette containing the standards and samples is inserted into the cassette loading bay section of the instrument. After the cassette is inserted and the protocol started, no further operator intervention is required.

The Pace-4® system library will store up to 20 protocols which are selected by the user to meet his laboratory's requirements. The library is protected from power interruptions and is permanently stored unless modified or erased through a sequence of operator commands.

If a sample cassette is being processed, the operator may load a second sample cassette into the loading bay and start a second assay. This cassette will be either held in the loading bay or transferred to a cooled storage area within the instrument. When the previous cassette has been processed, the second cassette will be retrieved and its processing begun automatically.

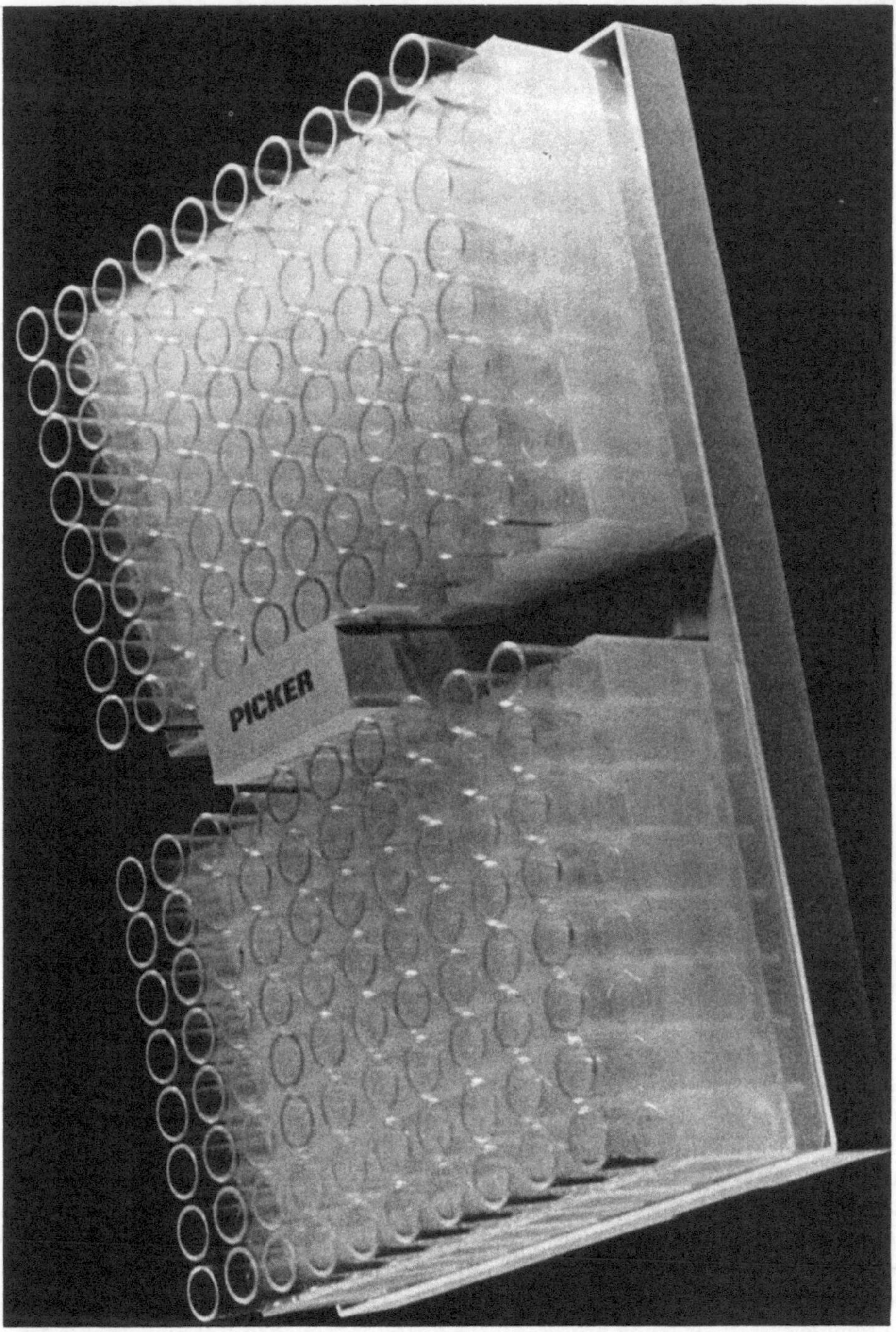

FIGURE 2. Pace-4® Sample Cassette.

A. Sample Transfer

When the protocol is started, the transfer of samples begins. The programmed volume of sample is transferred from the sample tube in the patient cassette to the reaction tube in the reaction cassette. The volume of sample transferred is determined by the assay protocol and discrete volumes between 25 $\mu \ell$ and 200 $\mu \ell$ can be accomodated by the system.

The Pace-4® system will also dilute samples as requested by the operator. If dilution is required, the patient cassette is prepared so that an empty tube for each dilution follows the sample to be diluted. The dilution ratio and location of the sample is programmed into the protocol for that assay. An exact aliquot of sample is aspirated and dispensed into the adjacent tube followed by buffer in the volume multiple required. Serial dilution of a single standard can also be performed in this manner.

If serum is to be transferred without dilution, then an excess volume of sample is aspirated to act as an interface in the probe. During each sample transfer, the tip of the probe is wiped to remove serum adhering to the outside of the probe. After the sample has been delivered to the reaction tube, the probe is flushed with buffer and wiped before the next sample is aspirated in order to eliminate the possibility of cross-contamination between samples.

B. Reagent Addition

The Pace-4® system reagent storage area will accomodate nine different reagents, so that several assays may be prepared for uninterrupted operation. Reagents are stored on the instrument in precision glass syringes within a cooled reagent chamber. Reagent delivery volume is controlled by the width of the syringe and the movement of the syringe piston through a calibrated stroke distance. Volumes between 50, 100, 200, and 500 $\mu \ell$ can be delivered at each dispense.

Since the syringes are positioned directly above the reaction tubes, no reagent is lost in delivery lines, and reagent waste is kept to a minimum. The large syringes will accomodate a mixing bar for those reagents that must be kept uniformly suspended.

The number of reagents to be used per assay, their volumes, and the times of reagent addition are all functions of the individual assay protocol. The Pace-4® reagent system is designed to accomodate a wide range of protocols. Each of these factors is programmed into the memory for each protocol and the system executes the assay as instructed. A diagram of the processing stages is illustrated in Figure 3.

C. Incubation

After reagent addition, the cassette is automatically transferred to an incubation chamber within the instrument. Two controlled temperature areas are provided for reaction incubation, and each will accomodate three reaction cassettes.

The incubation temperatures of the chambers may be set at 25° or 37°C at the discretion of the operator based on his individual assay requirements.

If the incubation time is very long, or at a temperature other than that provided, the operator may remove the cassette from the instrument for off-line incubation. To do this, the operator programs the instrument with the identification number of the cassette to be removed and the length of the off-line incubation period. The microprocessor will time the incubation period and alert the operator with an audible signal when the cassette should be returned to the Pace-4® system for processing. Once the cassette is returned to the Pace-4®, it will be processed to completion.

This option allows for unusual incubation temperatures and provides expansion capacity for laboratories assaying more than a thousand samples per day.

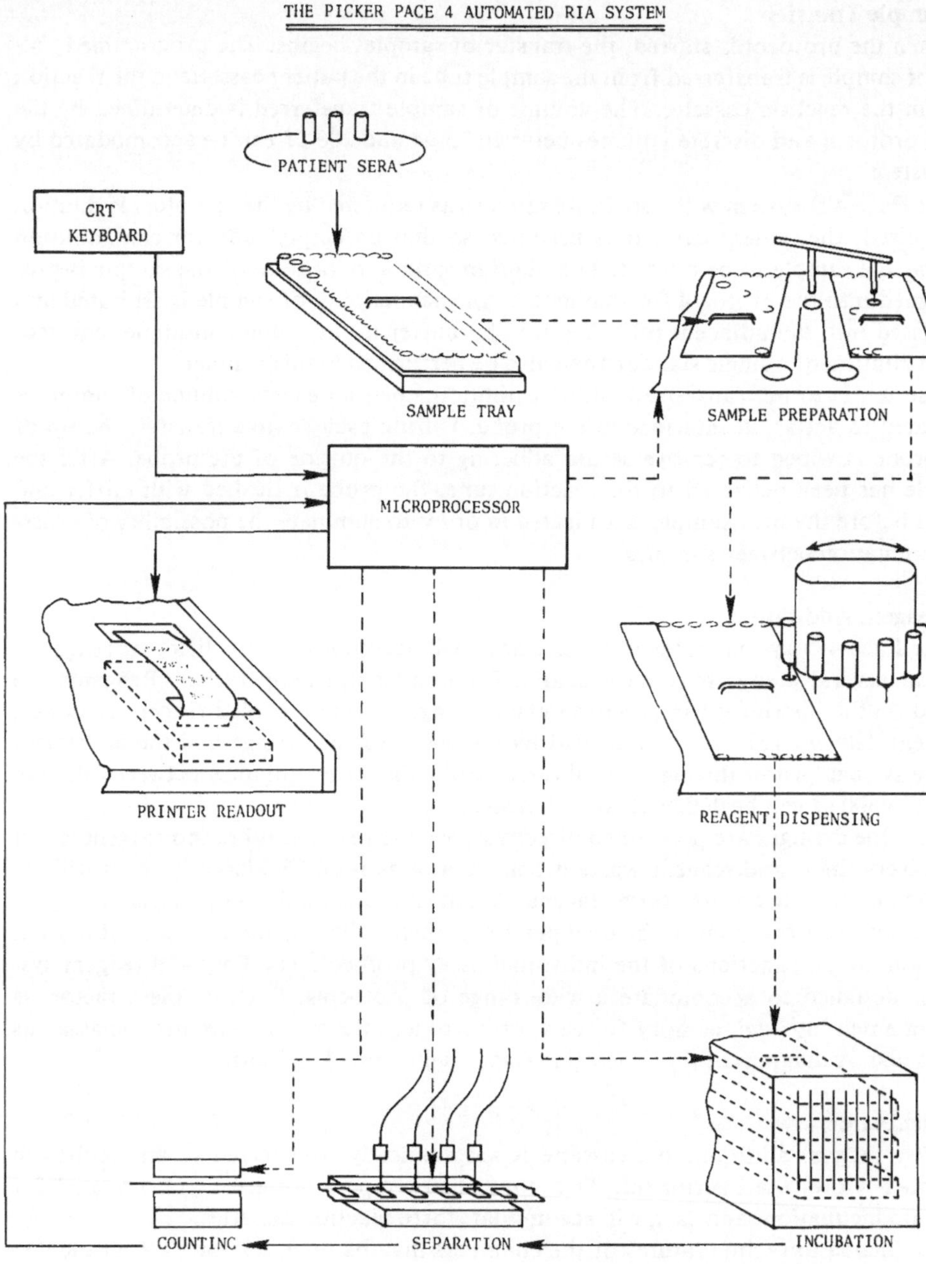

FIGURE 3. Diagram of the Pace-4® system illustrating the stages of sample processing.

D. Filter Separation

After incubation, the cassette is retrieved from the incubator and transported to the filtration section. The contents of four reaction tubes are aspirated simultaneously and passed through the filters. The particulates are trapped by the filter and the liquid passes through. The glass fiber filters will retain all particles larger than 0.7 μ in diameter.

The reaction tubes are then rinsed with buffer, and the buffer aspirated through the filter to complete the transfer of the solid phase and prevent the retention of any unbound antigen. The volume and number of washes necessary is a function of the type

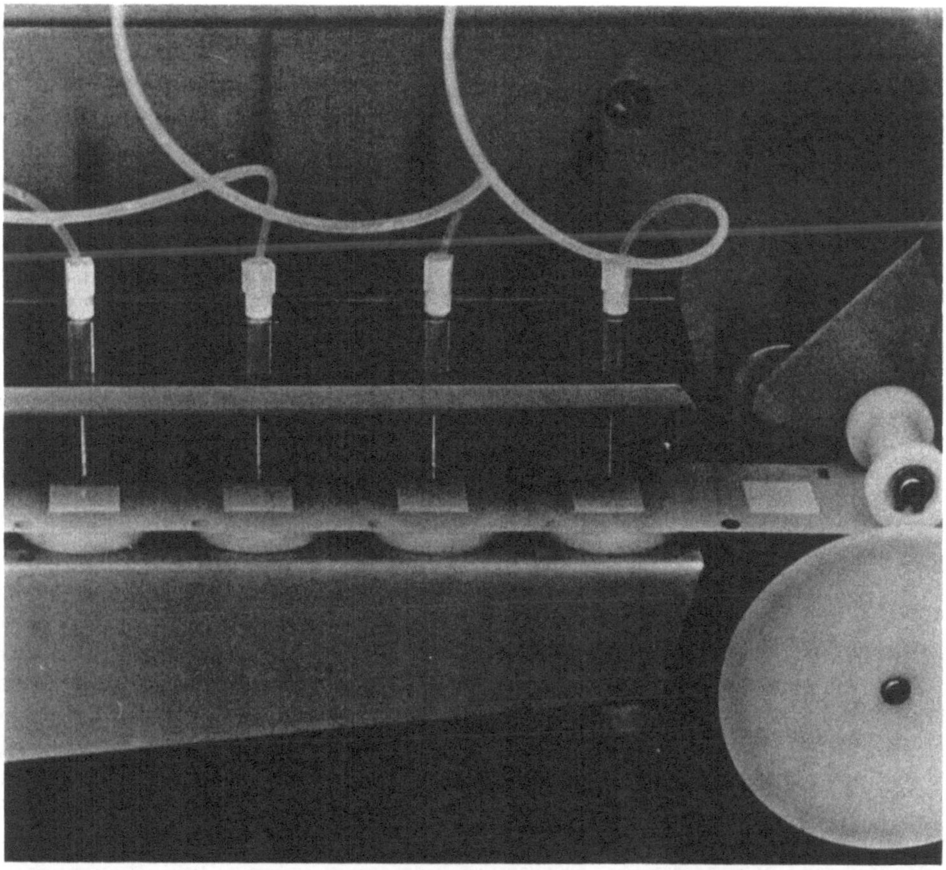

FIGURE 4. Filtration Section of the Pace-4® System.

of particle being transferred and is individually determined for each protocol. The filtration section of the Pace-4® system is illustrated in Figure 4. After the samples have been filtered, the filter strip is dried and sealed to prevent loss of particulates or contamination of the detectors.

E. Counting and Data Analysis

For measurement of the radioactivity trapped by the filter, the filter strip passes between a series of four pairs of opposed gamma radiation detectors. Each filter is counted by each detector pair for an equal fraction of the total count time. All pulses in the appropriate energy windows are summed from all eight detectors to determine the total count of the sample. The contents of four tubes are counted simultaneously, yet there is no effect of detector variation in the results since each filter is counted identically, and each sample count is the sum of all eight detectors. Both ^{125}I and ^{57}Co can be detected efficiently. These isotopes may also be counted simultaneously.

The count data is processed by the microprocessor according to the data reduction routine selected by the operator for the assay. A library of data reduction routines is provided, including CPM (no data reduction), ratio calculations, screen calculations, and four parameter logistic analysis for both single and dual isotope assays. The data reduction library may be expanded by the addition of new methods.

For those laboratories having auxillary data processing systems, the Pace-4® system contains RS232, IEEE 488, and current loop interfaces to transmit data directly to another system. The control program of the Pace-4® system provides the ability to review and edit the information before transfer.

The Pace-4® system is completely microprocessor controlled so that as many as four assays may be in process in different stages of the instrument at any given time.

IV. ASSAY PROTOCOLS

The Pace-4® system is designed to permit maximum flexibility for the adaptation of a wide range of available assays. Many manual assay procedures and commercially available kits can be accomodated without major modifications in procedure.

Although radioimmunoassay methods do vary substantially and not every method has been tested on the Pace-4® system at this time, excellent results have been obtained with double antibody procedures and solid phase procedures in which the antibody is bound to glass beads or microcellulose.

Assay protocols may be added to the protocol library by the user at any time for methods his laboratory may wish to adapt in addition to those protocols recommended by Picker.

A. Adaptation of Double Antibody Assays

Double antibody procedures have been adapted to the Pace-4® with minimal modifications. The Kallestad digoxin method is illustrative of the adaptation of this class of RIA procedure. The Kallestad digoxin procedure utilizes a preprecipitated double antibody reagent thus allowing the reaction to be completed in one 30 min incubation period. All reagent and sample volumes were within the flexibility of the Pace-4® system, and the assay protocol is adaptable to the Pace-4® system without modification. Experiments to reduce the volume of the reaction mixture by concentrating the double antibody reagent also gave excellent results.

Programmed parameters for the assay include serum volume (100 $\mu\ell$), volume of radiolabeled digoxin (50 $\mu\ell$), volume of antibody reagent (500 $\mu\ell$), incubation time (30 min), incubation temperature (25°C), data reduction type (4 parameter logistic), and volume of transfer wash buffer (2.5 cc).

Confirmation of the application of the Kallestad digoxin to the Pace-4® system included comparison results of standards and patient samples assayed by the Pace-4® system to the same standards and samples assayed manually according to the Kallestad published procedure. The results of the comparison studies are illustrated in Figures 5 and 6. The standard curves are parallel and the Pace-4® system curve shows a uniformly lower ratio of B/Bo. Experiments to determine the cause of the slightly lower curve indicate that the washing of the precipitate on the Pace-4® system filters removes completely the residual unbound antigen whereas in the manual procedure a slight volume of free antigen remains after the reaction tubes are centrifuged and the liquid decanted.

Comparison of 53 patient specimens covering the range of measurement of the digoxin assay yielded the results shown in Figure 6. Regression analysis of the resulting data pairs yielded a slope of 0.936, an intercept of 0.003 ng/mℓ and a correlation coefficient of 0.991. No outliers were observed.

B. Adaptation of Solid Phase Assays using Antibody Bound to Particles

Solid phase radioimmunoassays utilizing antibody bound to a particle have been adapted to the Pace-4® system. The Corning T_4 procedure is illustrative of the adaptation of this class of assays.

The Corning radioimmunoassays are developed with antibody immobilized to glass

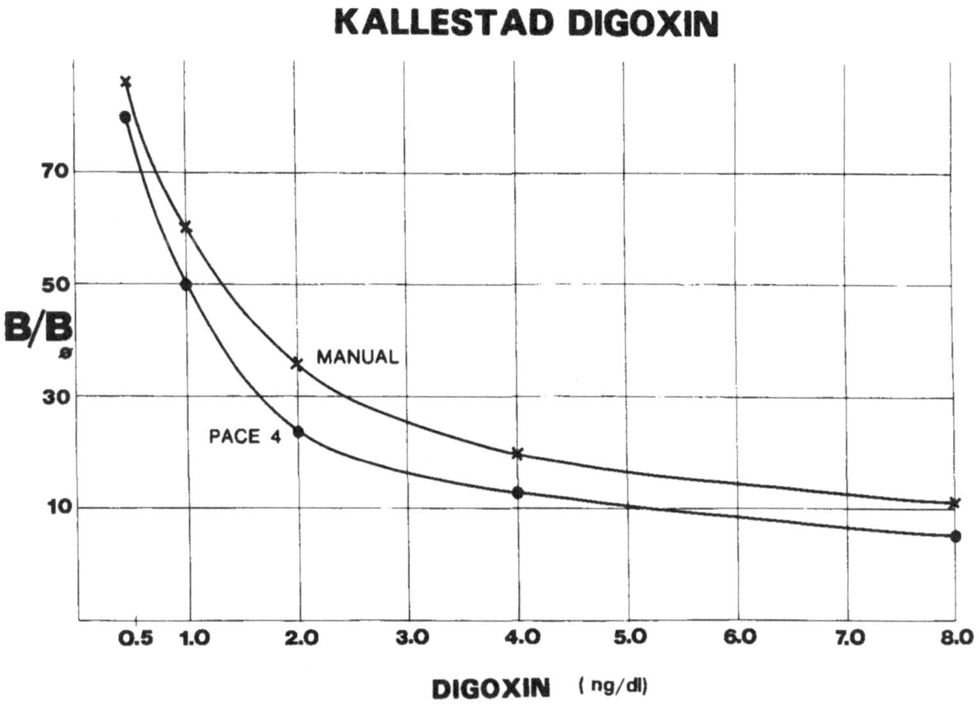

FIGURE 5. Comparison of digoxin standards processed by the Pace-4® System and the Kallestad manual procedure. Each point is the mean of duplicate determinations.

beads. The beads are approximately 1 μ in diameter and can be uniformly suspended in solution for assay. They can be completely filtered from solution on the Pace-4® system providing a clean separation of bound and free antigen.

The Corning thyroxine assay has been adapted to Pace-4® with only one modification in procedure. Programmed parameters for the assay include sample volume (25 μℓ), volume of radiolabeled thyroxine (100 μℓ), incubation time (1 hr), incubation temperature (25°C), data reduction type (4 parameter logistic), and volume of transfer wash buffer (7 mℓ). The Corning T_4 manual procedure requires the addition of 800 μℓ of antibody-bead slurry. This is a nonstandard volume for delivery on the Pace-4® system. In order to perform the procedure, the slurry was concentrated by allowing the slurry to settle in a container and removing the required volume of liquid from the top. Five hundred microliters of the concentrated slurry was dispensed per test on the Pace-4® system. Manual comparison assays were performed with unconcentrated antibody.

Results of the comparison studies are iilustrated in Figures 7 and 8. The standard curves are parallel and the Pace-4® standard curve shows slightly lower values as was also observed with the Kallestad procedure. Comparison of 51 patient samples covering the range of measurement of the thyroxine assay yielded the results shown in Figure 8. Regression analysis of the data pairs yielded a slope of 0.996. No outliers were observed. These results indicate comparable patient values between the manual method and the Pace-4® system.

Application of other assays of the same procedural types including Cortisol, TSH, and Free T_4 indicate that the Pace-4® system will accomodate a wide range of radioimmunoassays.

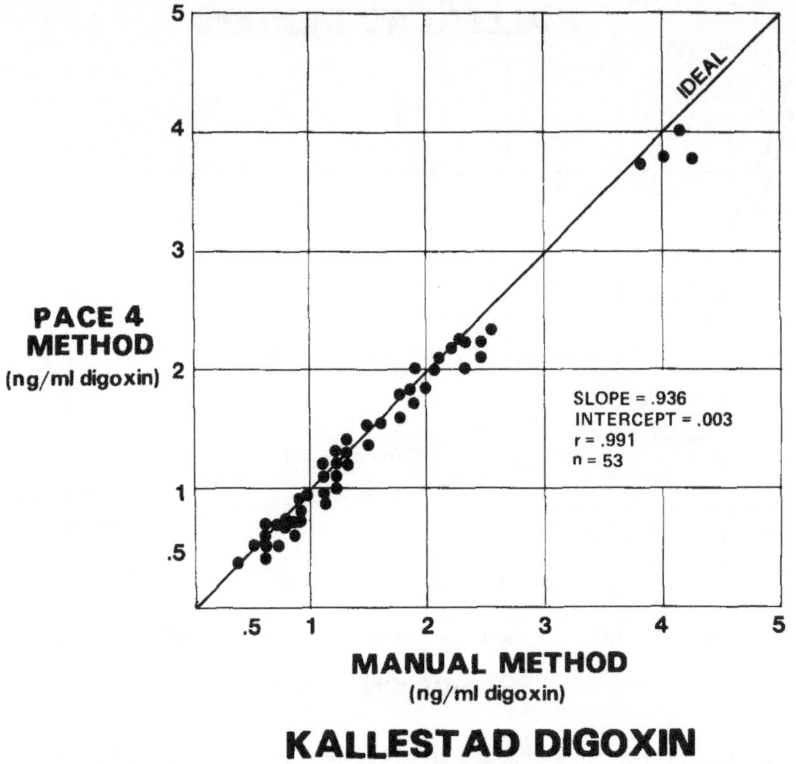

FIGURE 6. Comparison of 53 patient sera by the PACE-4 System and the Kallestad manual procedure.

V. DISCUSSION

The most significant advantage of the Pace-4® system is the flexibility to adapt a variety of radioimmunoassays to automation. The Pace-4® system has been developed with the specific objective of retaining the maximum degree of flexibility both in application of assay methods and in reagent supply sources while providing increased throughput and improved precision.

Each laboratory has its individual assay requirements both in variety and in volume. The flexibility of adapting a variety of available protocols allows the user to adapt those assays which meet his requirements and not be restricted in type or supply. New protocols may be adapted as the laboratory work load increases.

The benefits of automation are also evident in increased precision and speed and decreased total operating cost per result. The Pace-4® system is designed to capitalize on the cost benefits of bulk reagent supply and minimal operating expenses. Technician involvement is limited to assay start-up and routine maintenance, thus freeing highly trained personnel for newer laboratory procedures.

The Pace-4® system will process more than 1000 tests/day. The major limitation in assay turnaround is the incubation period for the antibody-antigen reaction since the samples are separated and counted at a rate of 220 tests/hr. This volume provides present utility and the capacity to expand as the laboratory work load increases.

The Pace-4® system has been designed to meet the needs of a variety of laboratory environments, both now and in the future.

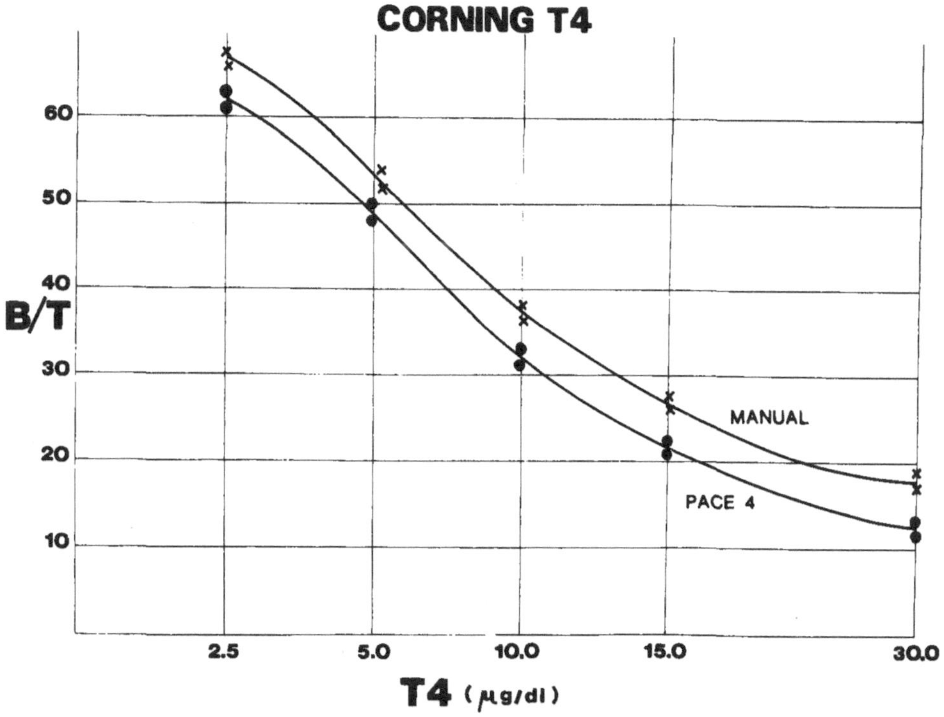

FIGURE 7. Comparison of thyroxine standards processed by the PACE-4® System and by the Corning manual procedure. Each point is the mean of duplicate determinations.

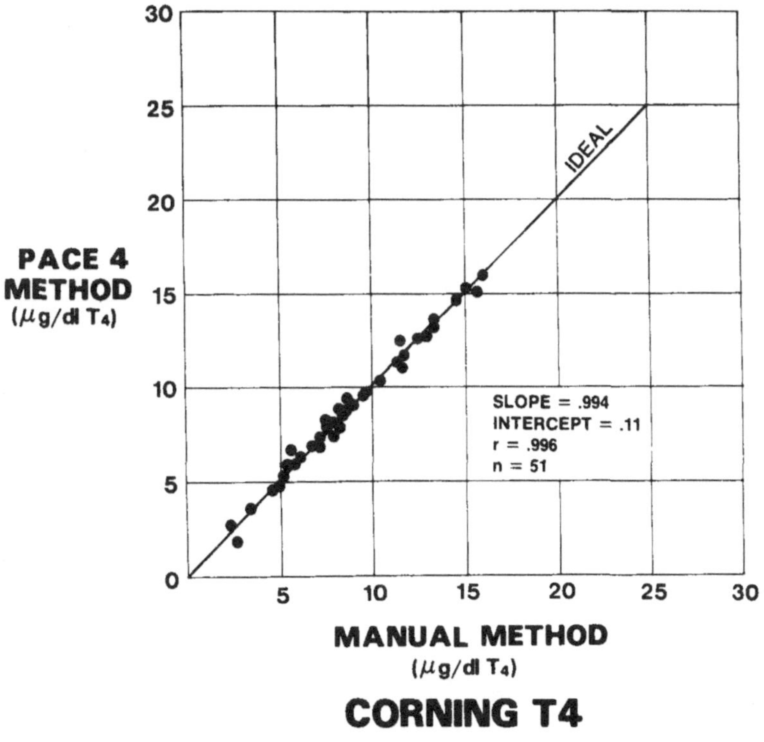

FIGURE 8. Comparison of 51 patient sera by the PACE-4 System and the Corning manual procedure.

ACKNOWLEDGMENTS

The advice and support of Professor Kenneth D. Bagshawe, Charing Cross Hospital, London, and Mr. James Kemble, Kemble Instrument Co., Sussex, who conceived and refined the technology through several generations is gratefully acknowledged.

Mr. Samuel C. Goldman, has contributed unfailing support to the progress of the Pace-4® system. Mr. Brian Wheaton, Mr. Blaise Lombardi, and Mr. Kreso Ukraincik have given generously of both their time and talent in the development of Pace-4® and their advice and cooperation in the preparation of the manuscript. Dr. Sidney Soloway contributed to the assay verification process.

To these people and many others who have worked to develop and refine the Pace-4® system, we extend our gratitude.

REFERENCE

1. **Bagshawe, K. D.,** Computer controlled automated radioimmunoassay, *Lab. Pract.,* Sept., 573, 1975.

Chapter 7

THE SYSTEMS COMPARED

L. Bowie

As indicated in the introduction to this text, the systems described in the previous chapters all serve to automate the various stages of sample preparation, incubation, separation, and quantitation. They all differ, however, in their overall design and analytical capabilities. Table 1 summarizes the most significant features of the individual systems.

As can be seen from the table, all of the various systems use dedicated gamma counters to detect radioactivity. This is a direct result of the widespread use of ^{125}I as a radioisotope in the preparation of labeled antigen. Since this gamma-emitting isotope has such useful properties (i.e., a combination of high specific activity which gives increased detection sensitivity along with a useful half-life of 60 days), it is even used to label noniodine containing substances. An additional advantage of using gamma-emitting isotopes is that radioactivity can be detected in solid or liquid samples without significant errors due to quench effects and without the requirement for volatile and expensive organic solvents necessary for beta detection. It should be noted that the Aria II® uses beta detection for the assay of the steroid hormones, cortisol, progesterone, and estriol. The reason for this relates to a theoretical improvement in antibody specificity if the ligand is labeled with ^{3}H (a beta emitter) as opposed to introducing iodine into the ligand by coupling it to ^{125}I iodinated tyrosine. Obviously the substitution of ^{3}H atoms for H atoms vs. the introduction of iodinated tyrosine results in labeled analogues which are more similar in binding properties with regard to the antibody or binder. In relation to this, it is noteworthy that the Aria II® is the only system which offers assays for progesterone and estriol.

It is not surprising that the most significant differences between the individual systems are the approaches used to separate bound and free fractions. No two systems use the same technique. The design of this portion of the various systems determines, to a large extent, the performance characteristics. For example, those systems in which samples are processed sequentially, one at a time, through the separation phase are the ones which in general have the lowest rate of analysis (e.g., Aria II®, Star®, Gammaflo™). On the other hand, the systems which have the most flexibility in choice of separation techniques allow for the greatest overall method flexibility (e.g., Pace-4®).

Although analytical throughput in terms of the number of analyses performed per hour varies from 30 to 240, the actual number of samples which can be processed in a given period of time by the various systems depends on a number of variables. The number of standards required per assay, the number of replicates per standard and how often standard curves must be rerun in a 24 hr period, all affect the throughput for actual samples. In addition, the "lag time" (the time between sample aspiration and the production of the first analytical result) as well as the size of the batch of samples which are processed in a given run contribute to the sample throughput. In general, however, the sample throughput is significantly lower than the number of analyses that can be performed per hour. For the systems described here, the sample throughput may vary from approximately 10 to 60 samples/hr assuming a batch of 30 samples and seven standards, all run in duplicate.

One of the major advantages of a totally automated system for radioimmunoassay

Table 1
COMPARISON OF AUTOMATED INSTRUMENTATION FOR RADIOIMMUNOASSAY

	Union Carbide Centria®	Micromedic® Concept-4®	Becton Dickinson Aria II®	Technicon Star®	Picker Pace-4®	Squibb Gammaflo™
Detection method	Gamma	Gamma	Gamma and beta	Gamma	Gamma	Gamma
Number of Detectors	3	2	1 each	1	4 pairs	1
Separation method	Centrifugal chromatography	Antibody-coated tubes	Antibody column	Antibody-coated magnetic particles	Glass fiber filter strip	Ion exchange column
Maximum thoughtput (analyses/hr, @ 60 sec/tube)	180	120	30	60	240	60
Estimated precision (CV, %)[a]						
intra-assay		7.4	2.1—5.4	<5	—	3.7
inter-assay		5.6—6.5	3.9—4.8	—	—	7.8
Sample volume (μl)	15, 35 or 50	2—200	20—500	90	10—250	10—35
Incubation Temp. (°C)	Ambient	Ambient, 37, 45	18—30	37	25 or 37	Ambient, 37
System cost ($)	39,500	65,000	68,000	45,000	79,000	60,000
Test available[b]	a,b,c,e,g,h,i,j	a,b,c,d,e,f,g,h	a,c,d,e,k,l	a,b,c,d,e,m	a,d,e,g,n	a,b,c,d,e
Adaptability to manual kits or special methods	Moderate (for nonsolid phase kits)	None	None	Limited	Great	Moderate (for nonsolid phase kits)

[a] Based on manufacturers' data for serum digoxin at a level of approximately 2ng/ml.

[b] Tests available on system as delivered: (a) Thyroxine (b) Triiodothyronine (c) T_3 uptake (d) Cortisol (e) Digoxin (f) Vitamin B_{12} (g) Insulin (h) Thyroid-stimulating hormone (i) Alpha fetoprotein (j) Ferritin (k) Estriol (l) Progesterone (m) Human placental lactogen (n) Free thyroxine.

is that the overall assay precision is significantly greater than that which could be accomplished manually. As can be seen from the table, the intra-assay coefficient of variation (for replicate analyses performed in the same batch) is for the most part less than 6%. I believe that it is safe to say that in most laboratories it would be very difficult to achieve a coefficient of variation (CV) of less than 10% using manual techniques for this kind of analysis. Inter-assay precision is somewhat higher, nevertheless the CV is usually less than 8 to 9%.

Capital investment costs also vary significantly from instrument to instrument. The overall expense involved in the operation of these systems must also include ancillary recurring costs such as those for reagents and expendable supplies. Although these costs also vary somewhat from system to system the potential user can expect the approximate cost per analysis to be $1.00 or slightly higher. For a somewhat more detailed analysis of both relative reagent cost per analysis and relative sample and analytical throughput on some of these systems, the reader is referred to Gochman and Bowie.*

The flexibility of the various systems to method modification and adaptability to new methods is also variable as is indicated in Table 1. In the case of Concept-4® and Aria II® the user is almost entirely dependent on the respective manufacturers as a source for reagents. On the other hand the Pace-4® can accommodate almost any of the current manual kit methods, with the exception of those employing antibody-coated tubes.

Another inherent advantage of these automated systems which is not as readily apparent is the possibility for performing nonequilibrium analyses. Since the incubation times can be reproduced with a high degree of accuracy/precision, times shorter than that required to achieve full equilibrium can be used without significant losses in reliability. This approach significantly shortens assay times and contributes to the increased rates of analyses on these automated systems. As a result, most of the manufacturers have taken advantage of this capability in the development of their recommended methods.

* Gochman, N. and Bowie, L. J., *Anal. Chem.*, 9, 1183A, 1977.

INDEX

A

B

C

D

O

P

Q

R

S

T

U

V

W

Y